AF674592

# THE STORY OF PLANET EARTH

teri

The Energy and Resources Institute

An imprint of The Energy and Resources Institute

© The Energy and Resources Institute, 2016

First published in 2016 by
The Energy and Resources Institute
TERI Press
Darbari Seth Block, IHC Complex, Lodhi Road, New Delhi 110 003, India
Tel. 2468 2100/4150 4900, Fax: 2468 2144/2468 2145
India +91 ▪ Delhi (0)11
Email: teripress@teri.res.in ▪ Website: http://bookstore.teri.res.in

ISBN 978-81-7993-535-4

All rights reserved. No part of this publication may be reproduced, stored in a retrieval system, or transmitted in any form or by any means, electronic, mechanical, photocopying, recording or otherwise, without the prior permission of the publisher.
All export rights for this book vest exclusively with The Energy and Resources Institute (TERI). Unauthorized export is a violation of terms of sale and is subject to legal action.
We would like to thank Richard Bizley for permitting us to use his images.
© Richard Bizley www.bizleyart.com

Author: Renu Anand
Publishing Head: Anupama Jauhry
Editorial and Production Teams: Nandita Bhardwaj, Rupak Ghosh,
Ekta Sharma, Himanshi Sharma; Aman Sachdeva
Design and Illustration Teams: Santosh Gautam, Vijay Nipane
Image Researcher: Shilpa Mohan

Printed and bound in India

This book is printed on recycled paper.

# CONTENTS

# INTRODUCTION

For thousands of years, man has wondered: How was the Earth formed? What is it made of? How old is it? How life began on it? Scientists have undertaken numerous investigations to find the answers. Today, owing to the wonders of modern technology, we know that the Earth is an evolving planet. From core to crust, it is undergoing continuous changes at an amazing pace. At the same time, we also realize how precious and special our planet is – the only known planet with oceans of water and an atmosphere rich in oxygen – as it sustains life. It would be imprudent not to learn from our planet's past, while addressing its dynamic present. But there's still so much more left to know about this blue marvel of the Universe.

## Evaluating the Earth's size

The Greek astronomer Eratosthenes of Cyrene (276–194 BCE) conducted the earliest documented experiment to measure the Earth's size. He noted that at noon on the summer solstice, the Sun was directly overhead in the town of Syene (Aswan) in Egypt. He observed a vertical post that didn't cast a shadow. In contrast, a similar vertical post did cast a slight shadow at the same time in the coastal city of Alexandria, located about 800 km to the northwest of Aswan. Correctly assuming that the Sun's rays are almost parallel when they reach the Earth, Eratosthenes calculated the Earth's circumference to be 250,000 stadia. The exact length of a stadia is unknown; however, it is believed that his measurement might have varied a little from the modern established value.

**ABBREVIATIONS**

**BYA** – Billion years ago
**MYA** – Million years ago
**KYA** – Thousand years ago
**YA** – Years ago
**BCE** – Before Common Era
**CE** – Common Era
*c.* – Circa

**GEOLOGIC TIME**

- The largest unit is **EON**
- An **EON** is subdivided into **ERAS**
- An **ERA** is divided into **PERIODS** and
- The more recent **PERIODS** into **EPOCHS**

## Legends about Earth's age

Since time immemorial, philosophers and thinkers have been making speculations about the age of Earth.

| | |
|---|---|
| **c.39,575 BCE** | Estimated to be the year of creation of Earth by the Egyptian priest Manetho, who lived possibly during 330 BCE |
| **c.39,000 BCE** | Believed to be the year of creation of Earth in ancient China |
| **3761 BCE** | Believed as the year of the world's creation by the earliest Hebrew thinkers |
| **3114 BCE** | The date set by the Mayans |
| **5500 BCE** | The date mentioned in the Holy Bible's Old Testament |
| **4000 BCE** | The date mentioned in the Holy Bible's New Testament |

## Scientific theories

Scholars proposed several scientific theories in the 18th and 19th centuries about the creation of Earth.

### 1

The "salt clock" method was suggested in 1715 by the English astronomer Edmund Halley, but it was flawed. The basic premise of the technique suggested the oceans did not contain any salt when they were formed and that any salt dissolved in the ocean stays there permanently. The first assumption can't be proved and the second one is incorrect, because elements of the ocean continuously leave and return via water cycle.

### 2

Every year, rivers and wind deposit an estimated 20 billion tonnes of sediment in the oceans. Of this, only one billion tonne of sediment goes away from the seafloor annually, according to Dr D R Humphreys, a physicist. So the current average rate of sediment on the ocean floor (400 m in thickness) would have taken about 12 million years to form. In short, he believes Earth can't be over 12 million years old, but his theory is flawed too. Different sediments with different compositions do not get dissolved in water at the same time, and the seafloor is constantly created and destroyed due to tectonic movements.

### 3

In 1799, William Smith, an English surveyor of canals, pointed out that the fossils of a particular kind were always found in the same layers (strata) of sedimentary rock. This method too did not help, as it only labelled rock layers as older, or younger, than others.

### 4

The British geologist Sir Charles Lyell (1797–1875) proposed that the same geological processes occurring today have operated since the beginning of time. While Lyell avoided quantifying the Earth's age, Charles Darwin (1809–82) speculated that 300 million years might be enough time to account for the geologic changes that Lyell had written about.

### 5

The British physicist Lord Kelvin (1824–1907), however, refuted Darwin's argument. He suggested that the Earth had originated as a hot molten body and that it was 20–400 million years old. Kelvin's estimate was flawed too, as he didn't take into account the heat produced by radioactive elements. Modern scientists opine that the Earth originated as a cloud of cool dusts and gases, warmed up due to the heat released by these radioactive elements.

### 6

With the help of precise dating techniques and instrument in the 20th century, the American chemist Bertram Boltwood (1870–1927) observed that the ratio between lead and uranium found in local radioactive rocks is usually highly uniform. This led him to conclude that lead is the final decay product of uranium. So he developed the radiometric dating method to estimate the age of rocks.

### 7

Using the uranium-lead and other, newer methods, geologists have dated a number of ancient rocks from Antarctica, South Africa, Australia, the Soviet Union, and North America. Their findings suggest that the oldest of these rocks were formed about 3.8 BYA. However, no rock has been identified as part of Earth's original crust. And, study of some meteorites or planetoids indicates the age of the Earth as being nearly 4.6 billion years.

# EARTH: A UNIQUE PLANET

Earth is the only known planet in our solar system that has large landmasses, meaning continents, and over a million islands. These landforms are surrounded by deep oceans and hundreds of thousands of streams and rivers. These continents, with huge plains, high mountains, deep valleys, ice caps, and deserts are largely covered with green vegetation. However, in the second half of the 20th century, we have caused changes beyond precedent to this precious, unique, and finite planet.

## The inhospitable planets

Other planets in our solar system are all very different from the Earth. Scientists studying the Earth and the rest of the cosmos feel that the chances of life occurring anywhere else in the solar system are remote. "So far, we haven't found it (life) anywhere else," says Alex Wolszczan of Pennsylvania State University, who co-discovered the first planets beyond our solar system. Mercury is seared by unbearably hot temperatures, due to its closeness to the Sun. The atmosphere of Venus, the brightest planet, is thick enough to trap the heat of the Sun, which gives it a blazing surface temperature of 471°C. Then comes Earth, beyond which exists Mars. Recent studies show signs that life might have existed on Mars, at least once. However, it is now a cold, waterless planet with a thin, dead atmosphere. Jupiter and Saturn, the gaseous giants of the solar system, are cold, hostile worlds. Some scientists believe that life might dwell beneath the frozen surface and wastelands of Europa, one of Jupiter's moons; however, this is mere speculation as yet. Uranus is so far from the Sun that its temperature remains below -200°C, making it the coldest planet of the solar system. Farthest from the Sun, Neptune is so distant that the Sun's light barely warms the planet. Also, its atmosphere, which includes methane gas, has the highest known wind speeds in the solar system with speeds near 1200 mph, which is faster than the speed of sound here on Earth, tearing at the planet's barren icy surface.

### Did You Know

NASA has confirmed the presence of water, although salty, on Mars. John Grunsfeld, astronaut and associate administrator of NASA's Science Mission Directorate in Washington, says, "Our quest on Mars has been to 'follow the water,' in our search for life in the Universe, and now we have convincing science that validates what we've long suspected."

A photograph of Mars

| PLANET | TEMPERATURE °C | | ATMOSPHERE |
|---|---|---|---|
| | MAX | MIN | |
| Mercury | 430° | -180° | very thin atmosphere of hydrogen, helium, and oxygen |
| Venus | 449° | 449° | thick carbon dioxide ($CO_2$), some nitrogen |
| Earth | 58° | -88° | nitrogen, oxygen, argon |
| Mars | 20° | -140° | $CO_2$ nitrogen, argon |
| Jupiter | -145° | -153° | hydrogen, helium |
| Saturn | -70° | -250° | hydrogen, helium |
| Uranus | | -224° | helium, methane |
| Neptune | -200° | -210° | hydrogen, helium, methane |

Our Solar System

## SIZE

Another "just-right" aspect of Earth is its size. If the size of our planet was much smaller, it wouldn't have been able to retain its precious atmosphere, and if it was much larger, it might have become a gas giant too hot for life.

## MOON

Scientists believe that we may also owe a debt to our nearest celestial neighbour, the Moon. It regulates our planet's rotation, preventing drastic movements of poles. This might have helped in stabilizing climate, which, in turn, caused life to form or evolve. Then the waxing and waning of Moon also causes tides in the oceans. This, according to scientists, might have made Earth the perfect place for early life to begin, evolve, and survive.

## WATER

Earth has water in liquid form on its surface, in an amount conducive to sustain life. The Earth is remarkable for it contains the right amount of water, not too much to cover the mountains, and not so little that it's a dry desert, as are Mars and Venus, our "sister" planets. Earth's water is also special in that it has remained liquid for so long. This is due to Earth's "just-right" location in the solar system. If it were closer, it would receive too much energy from the Sun, if it were farther out, it would have frozen.

## EARTH'S CRUST

Earth's crust is made up of different types of rocks, and it is composed of several rigid segments or tectonic plates that migrate across the surface over a period of millions of years. The slip-sliding movements of these plates have created the towering mountain ranges and deep oceans. Importantly, these tectonic plates allow for the carbon-silicate cycles to operate over geological timescales. This helps regulate temperature and keeps water in liquid state.

# THE BIRTH OF EARTH

In the beginning, there was no Earth and no Sun. Our Solar System, with the Sun in the centre and the planets with their accompanying moons, is a relative newcomer – 4.567 billion years old – in the 14-billion-year-old cosmos.

## THE BIG BANG THEORY

We live in one of the arms of Milky Way – a large spiral galaxy that is nearly 120,000 light-years across. It is home to 400 billion stars and our own Sun and solar system.

## Big Bang

All space and energy and matter came into existence in14 BYA, from a point smaller than the size of an atom, called singularity – a state of pure energy unbound by space and time. Probably a chance fluctuation in this atom provided an outlet for the enormous energy present in the pre-space-time state. Then, a tremendous explosion, called the Big Bang, occurred and in an instant there was not just something but everything that would ever be. This moment of creation of the Universe – the Big Bang – is the most mysterious and perplexing event in the history of the Universe.

When the Big Bang happened, the Universe was so hot that nothing solid could form, not even tiny particles of matter. Then from pure energy, less than one-thousandth of a second later, the first subatomic particles appeared. These were electrons and quarks, the essence of all the solids, liquids, and gases. This new Universe expanded in all directions from a size smaller than an atom to a fireball larger than a galaxy. Over billions of years, the Universe expanded and cooled. Stars, attracted to other stars by gravity, formed clusters of stars that formed gigantic spinning discs called galaxies. Galaxies are usually spiral-shaped – like our own Milky Way – or more elliptical, possibly formed by the merging of two galaxies or more. It is believed that the first galaxies formed around 13 BYA, and the Milky Way may be this old. Scattered debris of supernovas, subjected to the force of gravity, clumped together to seed new populations of stars by forming new nebulas, vast interstellar clouds of gas and dust. Each new nebula was more iron-rich and a little poorer in hydrogen than the one before.

## Birth of the Solar System

Our planet was a great nebula of gas and icy dust five billion years ago. Nine parts in ten of that cloud or nebula were hydrogen atoms; nine parts in ten of what remained were helium atoms. Ice and dust, rich in small organic molecules and microscopic mineral grains, accounted for the remaining one per cent. Then, almost 4.6 billion years ago, probably a supernova explosion from another star triggered the formation of our solar system. Slowly, the cloud began to rotate like a whirlpool. A large eddy at the centre contracted more rapidly than the rest of the cloud and over the course of a million years formed a dark, dense object, the "protosun". In the cloud

### Did You Know

This is a brown dwarf, surrounded by a swirling disc of celestial dust, which will be the elements of planet building. An artist's concept shows a brown dwarf surrounded by a swirling disc of celestial dust – essential for the building of planets. In simple words, it depicts a solar system in the making.

A brown dwarf

surrounding the protosun, certain atoms of gas combined to form compounds, such as water and ammonia. Slowly, solid dust crystals began to grow, as did metallic crystals, including iron and stony silicates. Gradually, gravitational and centrifugal forces flattened the spinning cloud into the shape of an enormous disc. Within this huge whirling disc, particles began to clump together. Similar bodies floated in space and collided with each other, sometimes merging to form boulders and smaller asteroids, sometimes breaking apart. While some of these clusters were torn apart in collisions, and others were broken up by the increasingly strong gravitational pull of the protosun, some merged into larger and larger clumps of matter, eventually forming boulders and smaller asteroids.

As these large bodies fought for survival, they used their power of gravity to pull in more and more matter from their surroundings into their mass. In this battle, some bodies gained matter, as others lost it. Ultimately, there were a series of large whirling discs in the region around the Sun. Each of these was a protoplanet, large enough to hold together under the strength of its own gravitational field. As each disc moved through space around the Sun, it acted as a sort of vacuum cleaner, sweeping up leftover material from the original cloud. Since the protosun was about a hundred times larger in diameter than the protoplanets, it exerted a much stronger gravitational pull that trapped light hydrogen atoms in its interior. It triggered thermonuclear fusion, releasing large amounts of energy. And it began to shine! So it became a star rather than a planet. The glow was fitful and dull red in the beginning, but later it became the golden yellow star that we see today. The growing bodies of matter closer to the Sun eventually formed the smaller, rocky, inner planets, Mercury, Venus, Earth, and Mars. Strong solar winds swept the nebular gases far away from the Sun. These eventually condensed into large dense gaseous bodies, or "gas giants", Jupiter, Saturn, Uranus, and Neptune – though they too contain rock at their core. Between the inner, rocky planets and the outer gaseous planets exist the main asteroid belt, and beyond the gaseous planets is the Kuiper belt of icy bodies.

## The creation of Earth

In the midst of all these cosmic activities, our future home was taking shape. Earth's temperature became high enough to melt the material at its centre as radioactive elements began to give off heat. Heavy metals – iron and nickel – that were spread throughout began to sink to form the molten core of the planet. During that time, Earth was a hot, blackened sphere, punctuated with glowing red cracks and towering volcanic magma as the molten rock frequently broke through fissures and came to the surface. As a result of incessant meteor impacts, vaporized rocks were blasted into orbit and the entire surface was turned into a molten, red-hot rocky slush. However, space is cold, and following each great impact, Earth's airless surface quickly cooled and blackened again.

### Did You Know

Crab nebula is an iconic supernova remnant in our Milky Way galaxy. Our planet was also a great nebula of gas and icy dust 5 BYA.

Crab Nebula

### Did You Know

A statue of Greek God Hades

Scientists named this period of the birth of Earth, lasting from 4.5–3.8 BYA, the Hadean Eon. after the Greek god Hades, who ruled the underworld, because during most of the Hadean Period, the surface of the Earth must have been like our image of Hell.

Molecules of hydrogen, water vapour, and other gases escaped from within to create an atmosphere above the planet's surface. But these light gases did not stay with Earth for long.

The Sun was already in action. Its rays were now striking the Earth with full intensity, breaking up the molecular compounds in its primitive atmosphere and scattering them into space. As a result, most of the atmospheric hydrogen and other light elements escaped from the Earth. This process eventually left behind a high concentration of the heavier, rarer elements of the Universe – elements essential for the formation of rocks, plants, and our own bodies.

## The birth of Moon

The origin of Moon remains an enigma to scientists. Did it form at the edge of the protoearth from the same cosmic matter, which produced the Earth? Or did it form elsewhere in space as a separate planet that was later captured by the Earth's gravitational field? Cosmologists favour the "big splash" theory, which says Moon is "Earth's child". According to this theory, Moon was born when a wandering planet, the size of Mars, crashed into the young Earth. Huge amounts of material were thrown into space, which eventually came together to form the Moon. This also establishes why Moon's rocks are very similar to those on the Earth.

# FORMATION OF THE CONTINENTS

The Earth was a very large, hot rock without a trace of water on its surface, about 4 BYA. Over a million years later, the Earth's surface, or crust, began to cool, creating a solid surface. Between 4.1 BYA and 3.8 BYA, a giant barrage of asteroids and comets pummelled the Earth, Moon, and the inner planets during an era, known as the Late Heavy Bombardment (LHB), leaving behind deep craters. Then erosion and movement of tectonic plates spread over a series of geologic periods, erased these craters from Earth's surface and formed continents. Tectonic movement, albeit very, very slow, is still drifting the continents apart.

## The oceans

Water on the young Earth came from two sources: outgassing from within the Earth and bombardment by comets. Gases are released from molten rock in the mantle of a planet by volcanic activity through outgassing. This was probably the primary source of gases for the early atmosphere. Comets and meteorites also bring gases which assimilate in the Earth's atmosphere.

The water on Earth stayed in gaseous form until the planet's surface cooled below 100°C. At this time, 3.8 BYA, water condensed into rain and poured onto the low-lying areas, which gradually became oceans. As water entered the oceans from the atmosphere, it brought with it dissolved gases. Water also flowed as runoff from the land, bringing in dissolved minerals from the rocks on the surface. These minerals include the salts which make seawater taste salty.

The geochemical cycles had their beginnings here. These cycles were well established about one billion years ago. Since then, the overall chemical composition of the oceans has been quite constant.

### THE BASICS OF PLATE TECTONICS

The Earth's interior constitutes of five layers – the inner core, outer core, lower mantle, upper mantle, and the crust. The lithosphere, which includes the crust and a part of the upper mantle, is rigid and rocky. Also, it extends to a depth of about 100 km below the Earth's surface. It is made up of about a dozen distinct blocks, called plates. Below the lithosphere lies the asthenosphere, which is so hot that its rocks are more flexible and supple than those above it. The tectonic plates float on top of the molten rock. When the continents and plates move it's called continental drift. When the floating plates spread apart, it's called a spreading centre. When they are moving together, it's called a subduction zone. When they are forced together, it is called a zone of convergence.

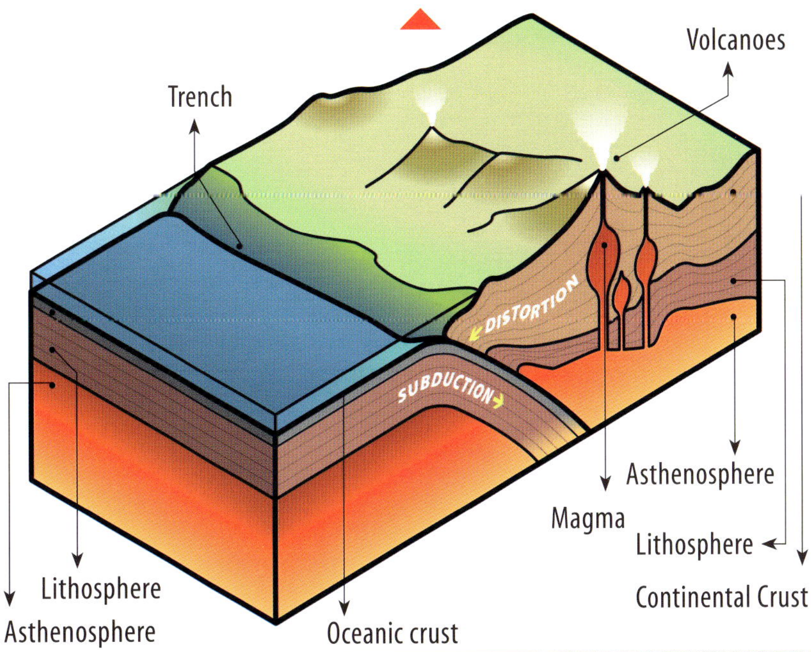

# Geologic **periods**

The geologic time on Earth, after the solidification of its crust, is broadly divided into four periods, namely, Eoarchean, Paleoarchean, Mesoarchean, and Neoarchean.

## EOARCHEAN (4–3.6 BYA)

Crust formation continued at this time. Due to the cessation of LHB, some of this crust survived and became a part of the continents, which formed much later. The Earth was mostly covered with water, with volcanoes and volcanic islands emerging here and there.

## PALEOARCHEAN (3.6–3.2 BYA)

During this 400-million-year long period within the Archean Eon, the areas of solid land expanded in size and the first continent-sized expanses of land appeared. These proto-continents no longer exist, but their remnants are sometimes found in cratons – areas of ancient rock – that survive on some of the continental shields today. Scientists are of the opinion that the first supercontinent Vaalbara came into existence in this era. This is based on the similarity in the sedimentary rock sequences on the South African Kaapvaal craton and the West Australian Pilbara craton (hence the name Vaal-bara). But this theory is controversial, and if Vaalbara did exist, it had started to break up by about 2.8 BYA.

## MESOARCHEAN (3.2–2.8 BYA)

This period covers the middle of the Archean Age. While continent formation continued, plate tectonics forced the separation of the ancient parts of South Africa and Australia. Another supercontinent which may have originated at this stage was the Ur, which is believed to have survived for a very long time, joining with other cratons to later form Rodinia and even later Pangaea.

## NEOARCHEAN (2.8–2.5 BYA)

The last 300 million years of the Archean Eon are called as the Neoarchean Period. During this period, large continents first appeared on the Earth, with modern plate tectonics.

*Outgassing of volatile gases through volcanic activity and bombardment by comets – factors responsible for bringing water on Earth*

© Richard Bizley; www.bizleyart.com

## Early continents

### RODINIA

About 1.2 BYA, fragments of continental crust came together due to the movement of the tectonic plates. They began to assemble into a giant continent, named *Rodinia*, a Russian word meaning "homeland". It is conjectured that rocks of ancestral North America, called "Laurentia", very likely formed its core.

For at least 350 million years, Rodinia was the major landmass of Earth. Like all the previous supercontinents, it was completely devoid of life, as all organisms still lived in the oceans. Its split caused new oceans to form and sea beds to rise, thereby moving animals and plants onto land.

### PANNOTIA

About 750 MYA, Rodinia broke into three continents: Proto-Laurasia, the continental craton of Congo, and Proto-Gondwana. These continents moved apart to form the Pannotia supercontinent. It broke up about 540 MYA and gave way to Pangaea, the last of Earth's supercontinents.

### PANGAEA

Pangaea was formed about 270 MYA and broke apart about 200 MYA. At this time, most of the dry land on Earth was one huge landmass that covered nearly a third of the planet's surface, and the surrounding ocean was known as Panthalassa. Earth's tectonic plates formed Pangaea and ultimately broke it apart.

## Present-day continents

**540 MYA** – The present-day continents of South America, Africa, Australia, and Antarctica, and the country India, formed the supercontinent of Gondwanaland. Present-day North America, Europe, and Asia combined to form another supercontinent called Laurasia.

**210 MYA** – The separation of Gondwanaland and Laurasia took place in the Triassic and Jurassic periods. Japan lay near the North Pole, India near the Antarctic. As a gigantic east-west crack appeared in the Earth's crust between the South American-African mass and Antarctica-Australia, India was liberated and started moving north. North America drifted northwest, the Eurasian plate twisted 20 degrees clockwise and moved off to the north. Africa rotated anti-clockwise, South America moved west, while Greenland broke off from Northern Europe. The last two continents to separate from one another were Antarctica and Australia, the latter migrating northwards to warmer climates.

**Did You Know**

In 50 million years, the Atlantic will be larger and the Pacific smaller. India will move farther east, Australia will almost touch Southeast Asia, East Africa and part of California will set out to sea, and new land will rise in the Caribbean Sea.

Present-day world

### FORMATION OF MOUNTAINS

Mountains are formed by slow, but continuous movement of Earth's crust over years. Most mountains form when the Earth's plates move towards each other, lifting the crust to a considerable extent.

As India hit Asia, the ocean that lay between them was pushed up into forming the Himalayas. Almost as violent were the happenings in the Mediterranean, where the African plate crashed into Europe, raising the mountain ranges of the Pyrenees, Apennines, and Alps. On the other side of the globe, the Americas, Asia, and Australia started moving towards one another across the vast Pacific, giving birth to the Andes, the Sierras, and the island arcs of the Aleutians and Japan.

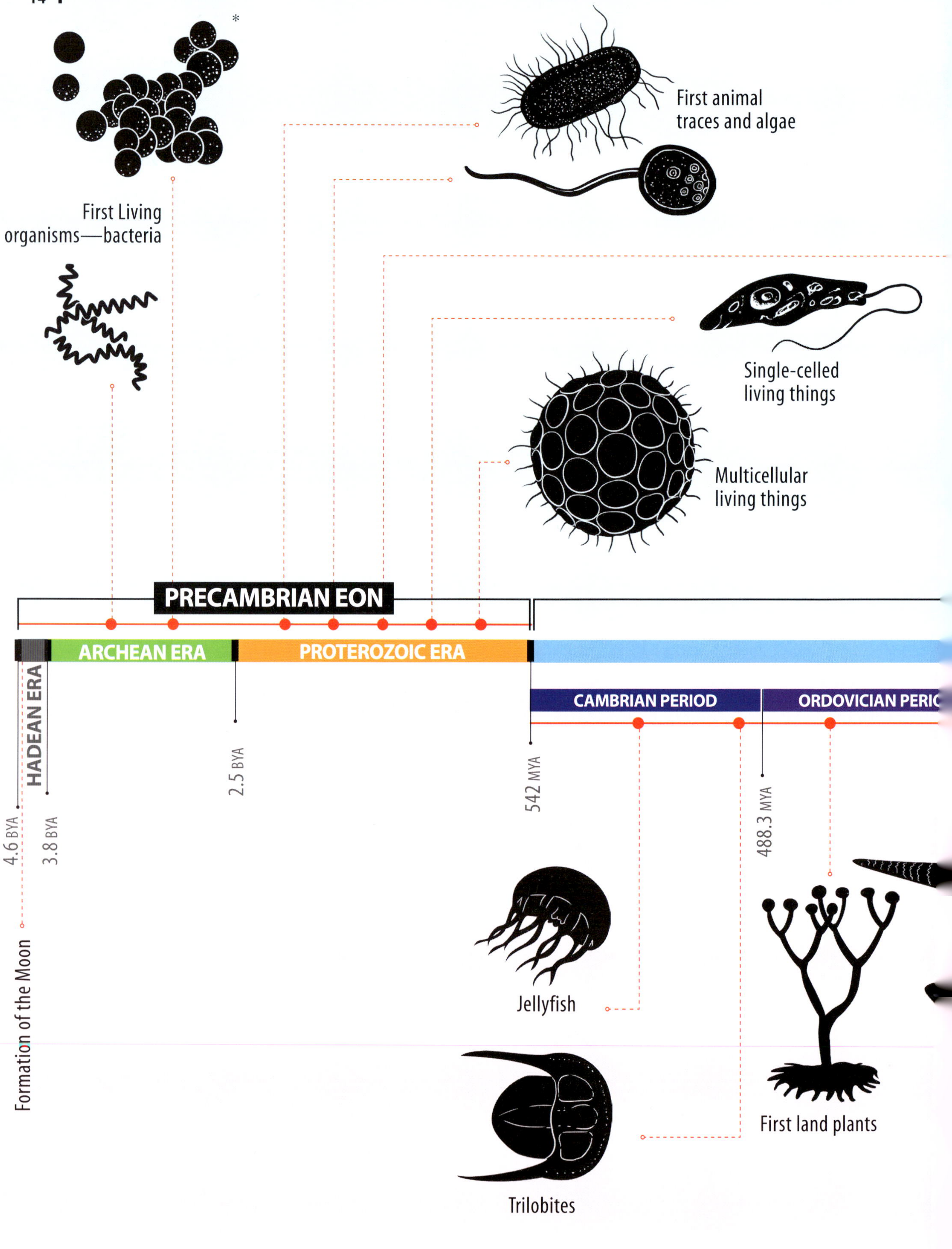


*The images shown here are representative life forms of the periods they refer to

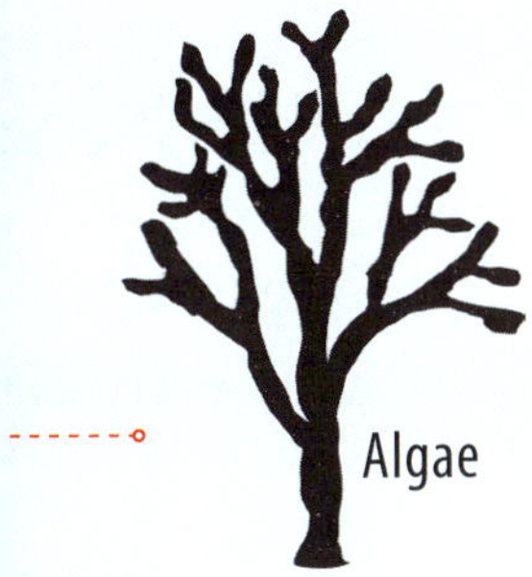


# GEOLOGIC TIME RANGE I

4600 MYA
Formation of the Moon

251 MYA
Permian Period

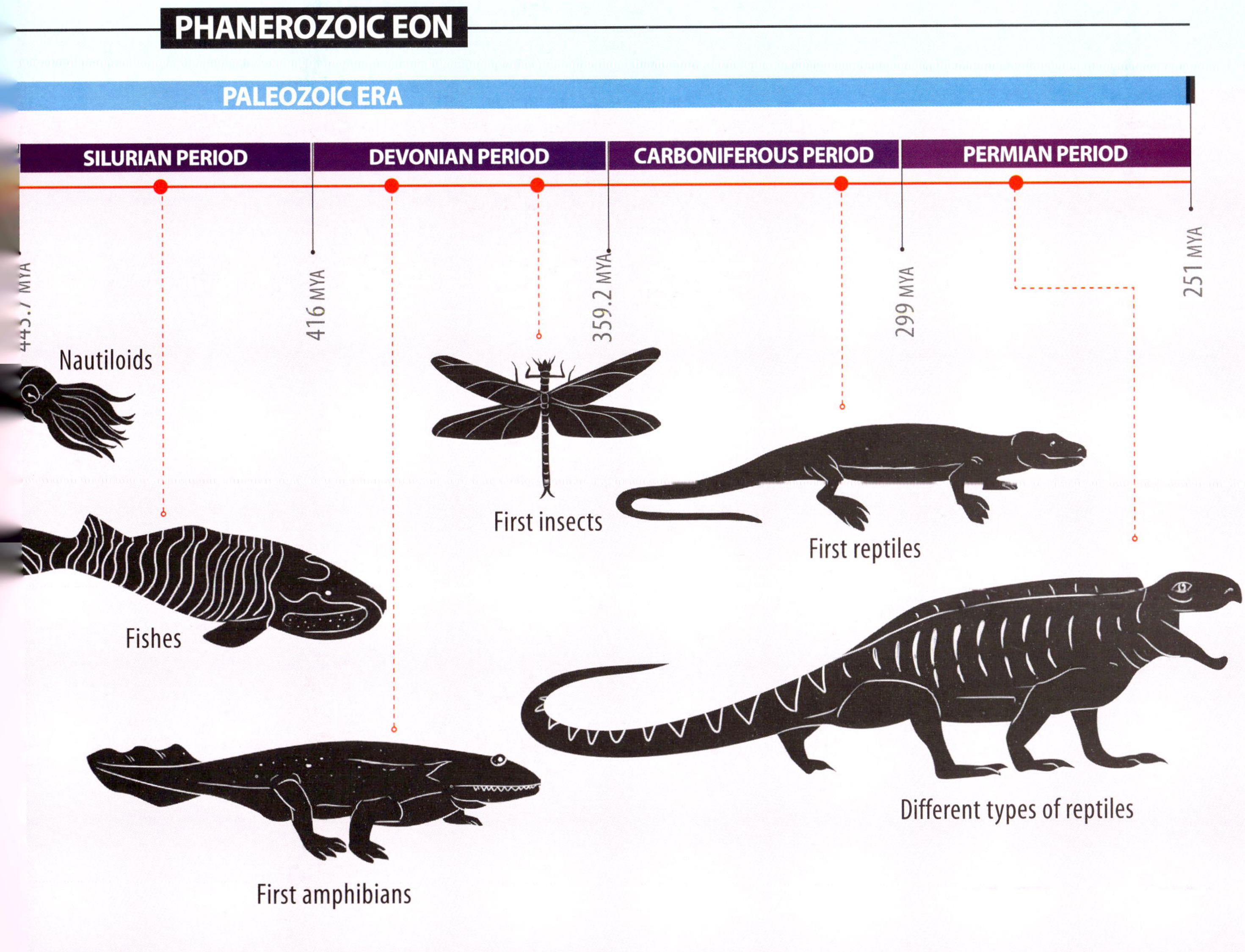

# THE BIRTH OF LIFE

Fossil bacteria in rocks, about 3.5 billion years old, suggest that life first stirred on our planet between 3.8 BYA and 2.5 BYA. This is called the Archaean Period, meaning the ancient period. At this time, the Earth was hotter and the atmosphere consisted mostly of a mixture of very poisonous gases, including carbon monoxide, carbon dioxide, nitrogen, hydrogen sulphide, methane, ammonia, and cyanide. Breathing this air would cause death to most life forms. At the same time, this mixture included many important elements that could be rearranged to produce amino acids (the building blocks of life). Then certain forces like lightning strikes appeared, which changed this barren wasteland into the beautiful garden that it is today.

## Primordial soup

Scientists had come up with the thought that life started in the ocean. Russian scientist Alexander Oparin and English geneticist John Haldane independently came up with the idea of a "primordial soup". They suggested that as the Earth's atmosphere was different during its initial years of existence, amino acids could be produced in the oceans by a series of chemical reactions of the gases present in the atmosphere with the help of energy present in lightning strikes. This concept came to be known as the "primordial soup".

The primordial soup consisted of basic components that are commonly found in life forms today. Over many millions of years, these ingredients of life formed the first living cells together, as they splashed around in the water, possibly by lightning strikes.

### Did You Know

In 1953, American scientists Stanley L Miller and Harold C Urey decided to test the primordial soup hypothesis. They put gases like methane, ammonia, and hydrogen alongside water in a closed apparatus to simulate the ancient atmosphere of the Earth. Next, they passed constant electric current through the system to simulate lightning strikes. After a week's time, they observed the formation of amino acids. Additionally, in 1961, it was found that hydrogen cyanide and ammonia in an aqueous solution could result in the formation of amino acids.

Primordial ecosystem

## Time zone

- Life on Earth began roughly around 3.5 BYA

## What was the Earth like

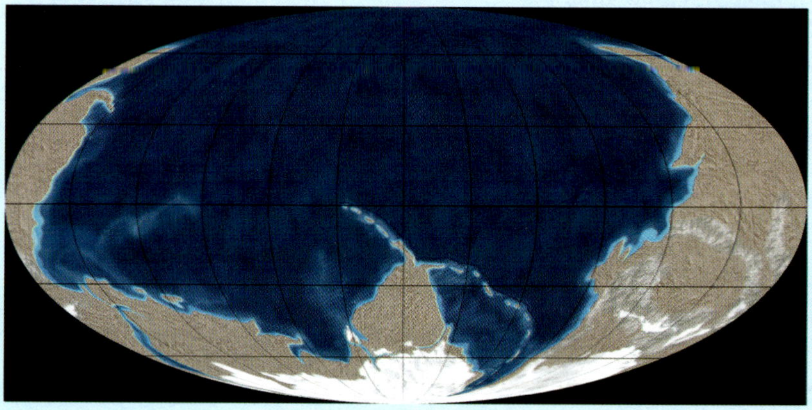

## Did You Know

Stromatolites are layered deposits, mainly of limestone, formed by the growth of cyanobacteria or blue-green algae (primitive one-celled organisms). Some of the first forms of life on Earth have been found in stromatolites. Stromatolites, up to almost 3.5 billion years old, have been found in the Pilbara region of western Australia.

Shark Bay's stromatolites

# Earliest life forms

In the beginning, and for a long period from then on, life comprised the simplest unicellular organisms, called prokaryotes, which lacked a membrane-bound nucleus. Bacteria and archaeans are prokaryotes. Then another form of life evolved, the eukaryotes that had a membrane-bound nucleus. Examples include fungi, plants, and animals.

### PROKARYOTES AND EUKARYOTES

Prokaryotes were early forms of life. These single-celled organisms were of two types. The first type fed on the proteins, amino acids, and lipids present in the water. For example, bacteria. In contrast, the eukaryotes like cyanobacteria, or blue-green algae, had the ability to create their own food using the power of sunlight, so they were called producers. They took in $CO_2$ and water from their environment and used these materials to make carbohydrates and energy and released oxygen as a by-product. Over billions of years, they increased the amount of oxygen and decreased that of $CO_2$ in the air, making it easier for other organisms to evolve. This led to the emergence of larger and more complex multicellular life forms in the oceans.

# Ediacaran period

Ediacaran period (635–542 MYA) marked the final stages of the Precambrian Period. Some of the creatures of this period, which required oxygen and were precursors of organisms with skeletons, were similar to jellyfish, sea anemones, and soft corals of today, while others bore little resemblance to modern life forms. All of them lacked bones, shells, teeth, or other hard bony parts. While the origin of the Ediacara fauna resulted from a rise in the oxygen level in the atmosphere, they became extinct about 542 MYA possibly due to heavy grazing by early skeletal animals.

## Fossil Fact

The first animals with more than one type of cell (metazoans) appeared over 600 MYA. The earliest known multicellular animal is the Ediacara fauna – an assemblage of soft-bodied organisms whose fossils were found in 1946 by a geologist named Reginald C Sprigg in the Ediacara Hills of South Australia.

*Dickinsonia costata* fossil

*Green algae on rocky North Irish coastline*

# CAMBRIAN **PERIOD**

The Ediacaran Period was followed by the Cambrian Period. It witnessed a never-before and never-again "explosion" of diverse life forms, including the first animals with a hard outer skeleton. Life was still evolving only in the sea. This Cambrian Explosion witnessed hard-bodied animals that easily fossilized. The climate was warmer than what it is today (22°C as compared to 14°C today). It is thought the final stages of Precambrian time were marked by a prolonged global ice age, which may have led to widespread extinctions.

## Time **zone**

- Began 542 MYA
- Ended 488.3 MYA

## What was the **Earth like**

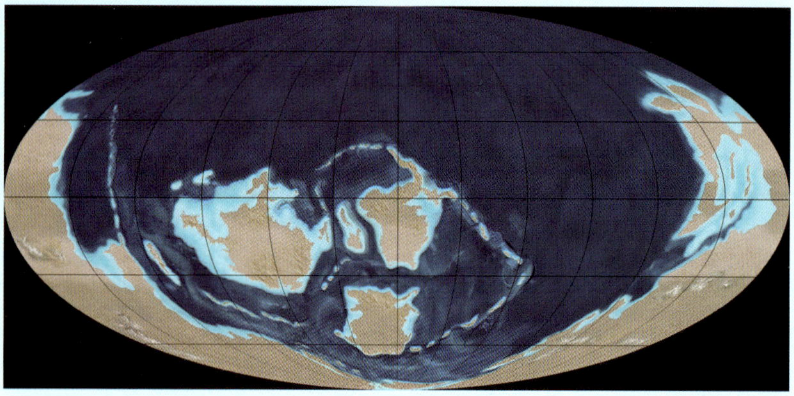

## What **lived then**

### EARLY INVERTEBRATES

Life in the oceans was quite plentiful during this period. The earliest living organisms were invertebrates that did not have a backbone. Many strange types of invertebrate animals lived in the Cambrian seas like the soft-bodied sponges without organs or a nervous system. They drew water through their bodies and filtered out food particles.

Cnidarians, which included sea anemones, corals, and jellyfish, had sac-like bodies and a simple digestive system with a mouth but no anus. They caught food using tentacles. Annelids, or segmented flatworms, had fluid-filled body cavities and they breathed through their skin.

Sea lilies were also common. They had long stalks and arms, with sticky tentacles to catch food. They were related to the present-day sea urchins and starfish.

### EARLIEST VERTEBRATES

Vertebrates are animals that have a backbone or spine. However, this characteristic in Cambrian fish was yet to be truly defined.

The most famous of these, *Pikaia*, looked more like a worm than a fish, but it had four features crucial to later fish and other vertebrates viz. head distinct from its tail, two-sided symmetry, V-shaped muscles, and most importantly, a nerve cord running down the length of its body. Because this cord wasn't protected by a tube of bone or cartilage, *Pikaia* was technically a "chordate" rather than a vertebrate, but it still lay at the root of the vertebrate family tree.

Two other Cambrian proto-fish were slightly more able-bodied than *Pikaia*. *Haikouichthys*, which also did not have a calcified backbone, is considered to be the earliest jawless fish. This inch-long creature had rudimentary fins running along the top and bottom of its body.

*Myllokunmingia* was slightly less elongated than either *Pikaia* or *Haikouichthys*, and it also had pouched gills and possibly a skull

Anemones

Orange tree coral

Light jellyfish

Sponge coral

## Fossil Fact

*Trilobite Paradoxides* are extinct marine invertebrate animals that evolved hard parts, such as shells, which readily fossilized and left a clear record behind.

Trilobite Paradoxides

made of cartilage. Other fish-like creatures may have predated these three genera by tens of millions of years; unfortunately, they haven't left any fossil remains.

## Cambrian mass extinction

Just when the Earth's oceans were teeming with the early complex animals, many of them were wiped out. A study of 500-million-year-old rocks shows high levels of sulphur and carbon – indicating that significant amounts of these elements were being buried in the seas. As the oxygen level fell and a combination of hydrogen and sulphur resulted in the formation of poisonous hydrogen sulphide, many Cambrian species were killed off.

Rocks in some parts of the world belonging to the period dating from around 500 MYA also show a band of deposited iridium. This rare element is associated with intense volcanic activity. Probably volcanic eruptions and dropping oxygen levels were devastating to life in the oceans at the time, thereby leading to mass extinction of many genera.

## Did You Know

Sea-scorpions were the largest invertebrates to have ever lived on Earth. One type was two metres long and used its powerful pincers to crack open the shells of the early fish and snails.

*Jaekelopterus*

*The first fish-like creatures were similar to a coral-like sea squirt, whose larvae resemble primitive fish*

© Richard Bizley; www.bizleyart.com

# ORDOVICIAN **PERIOD**

In the Ordovician Period, the Earth underwent a lot of changes. Life remained confined to the seas, with new animals evolving in place of those that succumbed to the Cambrian extinction. The first primitive plants began to appear. Pannotia – the supercontinent formed about 600 MYA, when the three pieces into which Rodinia had broken up came back together again – fragmented into several small parts. The largest piece Gondwana comprised present-day China, India, Africa, South America, Arabia, Madagascar, Australia, and Antarctica. Throughout the Ordovician Period, Gondwana drifted south, eventually settling on the South Pole (which was located where north-west Africa stands today).

The Earth's climate was mostly warm and wet, while sea levels were as much as 1,970 ft higher than what it is today. Due to frequent volcanic eruptions, $CO_2$ levels in the atmosphere were 14–16 times higher than the current levels.

But once Gondwana stopped drifting, massive glaciers formed over the supercontinent's centre. This marked the beginning of a 20-million-year-long ice age, leading to a mass extinction.

## Time **zone**

- Began 488.3 MYA
- Ended 443.7 MYA

## What was the **Earth like**

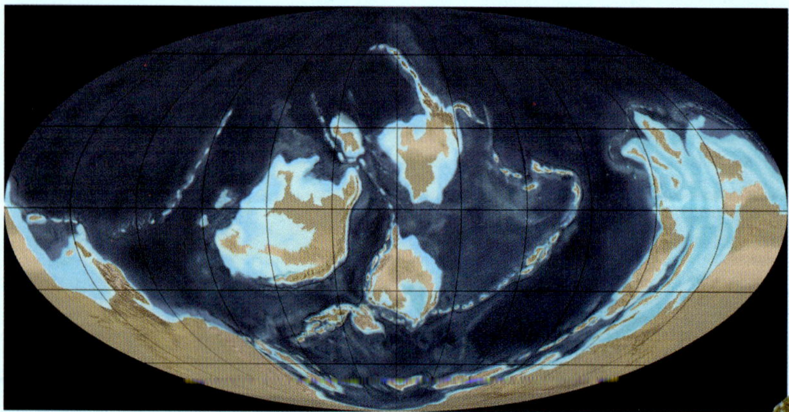

## What lived then

### INVERTEBRATES

A rich variety of marine life flourished in the seas during the Ordovician Period. New species took birth in place of those who were lost during the Cambrian mass extinction. Nautiloids and conodonts are some examples.

Nautiloids, a type of mollusc that had a squid-like appearance, existed during this period. Their grasping tentacles made them expert swimmers and predators.

*Conodonts* were small marine creatures (0.2–6 mm in length). They had fins and large eyes and resembled eels. *Conodonts* left behind tiny fossil teeth that led to their identification, and it is believed that they had a backbone.

Corals continued to survive, but the reefs were predominated by algae and sponges.

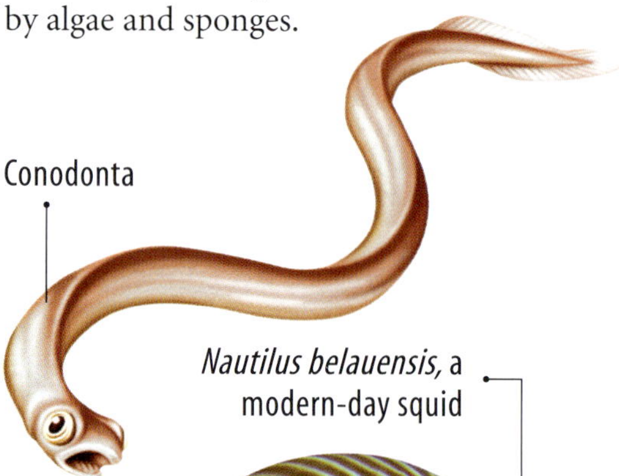

Conodonta

*Nautilus belauensis,* a modern-day squid

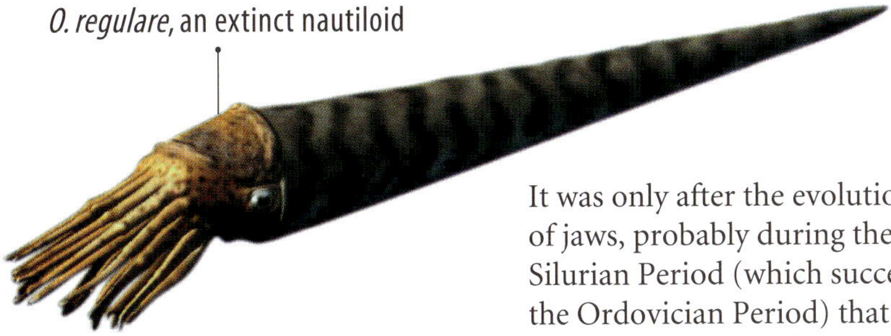

*O. regulare*, an extinct nautiloid

## VERTEBRATES EVOLVE

Jawless fish (*agnatha*) were the first true vertebrates that were sturdy. These fish had downward-pointing, jawless mouths, with which they sucked, scraped and filtered food off the ocean floor. A unique feature of these fish was the presence of tough, bony shields on the front of their bodies, which protected them from predators like giant sea scorpions and other large arthropods. However, this armour plating made them slow and weak swimmers.

The most notable jawless fish of this period were the six-inch long finless *Astraspis* and *Arandaspis*. Resembling giant tadpoles, their large heads were covered by armour while plate-like scales covered the tail. These fish lived in shallow waters, coming up to the surface to gobble tiny animals with their slitlike mouths.

It was only after the evolution of jaws, probably during the Silurian Period (which succeeded the Ordovician Period) that fish began to dominate the oceans. With the evolution of fish with jaws, the jawless species began to decline. Today, there are only two groups (lamprey and hagfish) and about 80-odd living species of jawless fish left in the world.

### Fossil Fact

A partially uncoiled Nautiloid of the genus *trilacinoceras* from the Hunan Province, China. The specimen measures 9.5 cm in length.

Fossil of a partially uncoiled nautiloid

Arandaspis

Astraspis

### Did You Know

During this period, certain species started looking for opportunities on land. The horseshoe crab left its saltwater habitat and moved into freshwater and shallow lagoons. Also, the first plants appeared on the previously barren land.

Horseshoe crabs

## Ordovician mass extinction

Diversity of life during this period was cut short by the Ordovician-Silurian extinction - a mass extinction which wiped out an estimated 85 per cent of all species. It was caused by rapid lowering of temperature and fall in sea levels, both of which occurred as a result of glaciation and the onset of an Ice Age. All the extinctions occurred in the oceans, since there were no terrestrial plants and animals yet.

# SILURIAN PERIOD

In the Silurian Period, rising temperatures caused the large glacial formations from the Ordovician Age to melt, resulting in dramatic rise in sea levels. Consequently, there occurred a re-emergence of the earlier eras' marine environment. Unlike the previous eras, when atmospheric fluctuations were common, this period was marked by a relative stabilization in the Earth's climate. A major feature of the period was the emergence of mountain ranges in eastern North America and north-western Europe. However, the most unusual and distinctive aspect of the period was a combination of low continental elevation and much higher sea levels, as compared to today. Towards the end of the period a new supercontinent, called Euramerica, was formed.

*Prehistoric underwater life during Silurian Period*

## Time zone

- Began 443.7 MYA
- Ended 416 MYA

## What was the Earth like

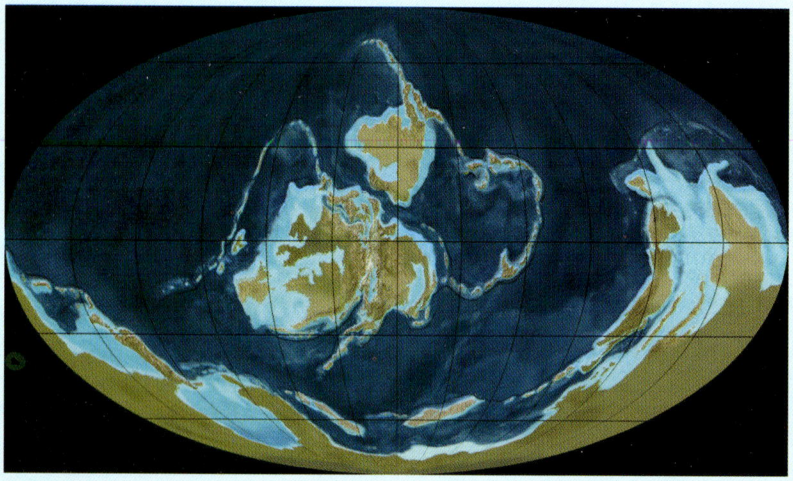

## What lived then

### VERTEBRATES

Freshwater fish and fish with jaws appeared in the sea. A group of fish called acanthodians, or spiny sharks, were the first true-jawed fish. While only about a foot long, with a small head and large eyes, these fish were soon to become one of the planet's foremost predators.

Acanthodian fish

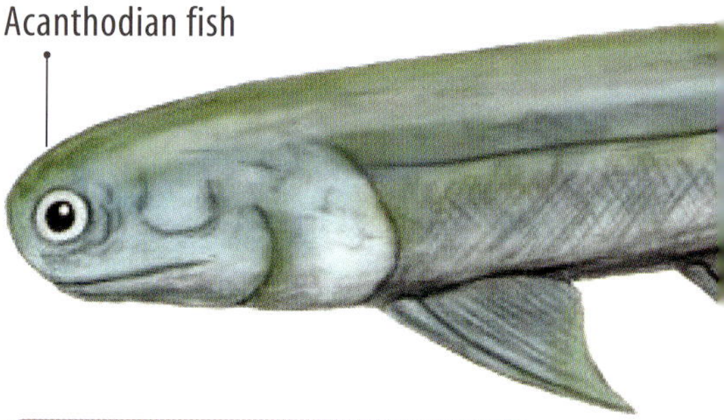

## Fossil Fact

This 400 million year old fossil of *Birkenia elegans* is a beautifully preserved example of one of the earliest-ever Acanthodian fish. It was jawless and covered in armour plating. Another noticeable feature was its upside-down tail.

A fossil of *Birkenia elegans*

© Richard Bizley; www.bizleyart.com

An artist's impression of a giant sea scorpion

## INVERTEBRATES

In stark contrast were the giant sea scorpions. Sometimes growing up to 6 ft in length, these terrifying creatures with powerful, claw-like pincers were the Earth's largest-ever arthropods.

Corals evolved into reef ecosystems massively and dominated the seas. These reefs were home to a wide variety of corals and tentacled microscopic sponges called stromatoporoids. Widespread availability of tiny planktonic creatures in the oceans helped the growth of corals and other marine organisms.

## Silurian mass extinction

This period came to an end about 416 MYA with the occurrence of three small-scale extinction events, relatively minor compared with those of other geologic periods, which were triggered by climate change.

### Did You Know

The earliest known fossils that offer the first evidence of life on land – both terrestrial animals and plants – are from this period. The fossils include those of a group of fungi, as well as those belonging to the first centipedes, millipedes, and ancestors of spiders.

Land plants made their first appearance along the edges of rivers and lakes. Most of these plants belonged to the genus *Cooksonia,* which is commonly considered the oldest-known terrestrial plant. These early plants bore no leaves, had firm stems, and were only about 6 cm tall. Their presence significantly increased the proportion of oxygen present in the atmosphere. Over time, mosses and other plants evolved, providing a thin layer of vegetation adjacent to water bodies. This development led several marine animals to shift to land.

A cartoon of *Cooksonia*

# DEVONIAN PERIOD

This period is named after the county of Devon, England, where old red sandstone, produced when North America collided with Europe, was first studied. At this time, our planet was changing its appearance – the Gondwana supercontinent was moving steadily towards the north, while another supercontinent Laurasia was created around the Equator.

The Devonian Period is also known as the "Age of Fishes", as the period is noted for wide and rapid rise and diversification in fish in both fresh and marine waters. The appearance of a wide variety of life forms – such as ferns and plants with true roots and leaves – is often referred to as the "Devonian Explosion".

## Time **zone**

- Began 416 MYA
- Ended 359.2 MYA

## What was the **Earth like**

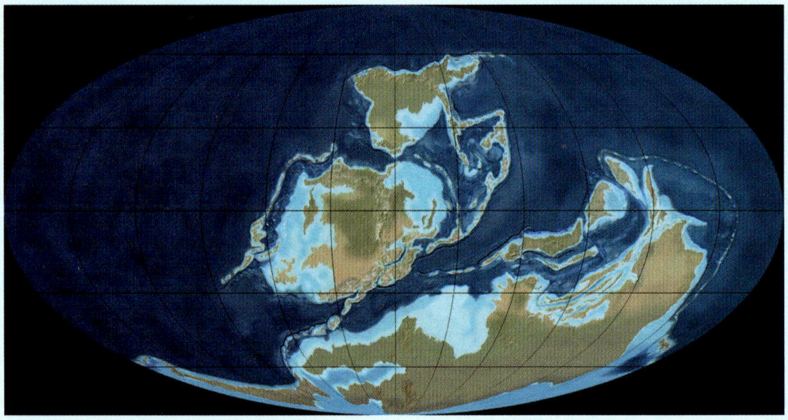

## What lived **then**

### LIFE ON LAND

Invertebrates of this period were mostly similar to those found during the Ordovician Period. The plant-covered land made a good home for mites, wingless insects, and spiders. Towards the end of the period, a primitive four-legged vertebrate, tetrapod, developed the ability to live outside water and move on land.

### THE AGE OF FISHES

Thousands of species of fish developed in the seas. Besides jaws, the fish started developing gills and paired fins as well.

The first fish with all three were the placoderms. Their evolution veered off in two directions. One group, comprising armoured jawed fish – ancestors of sharks and rays – became extinct during the mass extinction at the end of the Devonian Period.

The other branch constituted shark-like fish (chondrichthians) and bony fish (osteichthians). The former had skeletons made of cartilage, while the latter had bodies covered in scales. Scientists feel that the bony fish are ancestors of all four-limbed terrestrial vertebrates, including dinosaurs and mammals, as their fins later evolved into "legs".

### THE REEF BUILDERS

The seas were dominated by clam-like marine animals, called brachiopods, and corals that built some of the largest reefs the world has ever seen in shallow waters. From the shells of the now-extinct ammonoids, which were ancestors

## Fossil Fact

*Dipterus valenciennesi*
Size: 125mm

Fossil Site: Achanarras Slate Quarry, Caithness, Scotland

Dipterus, one of the earliest-known lungfish that had primitive lungs as well as gills made its first appearance during this period. They were able to survive long periods of dry conditions by gulping air with their lungs, and then reverting to an aquatic, gill-powered lifestyle when rivers and lakes they lived in filled up again with water.

*Dipterus* fish fossil

## Did You Know

In 2004, the discovery of the fossil of a fish with a flattened head like that of a crocodile provided the transition between fish and tetrapods. The first four-legged vertebrate to walk on land, called the **Tiktaalik**, had fins, scales, and gills like most fish, and wrist bones, flat head, neck, shoulder, and thick ribs.

Tiktaalik

*An early Devonian landscape of plants growing on land. Most vascular plants were about 2 ft high*

© Richard Bizley; www.bizleyart.com

of squids and octopuses, were formed large limestone deposits.

## What grew then

### SPRAWLING PLANT LIFE

Since both the supercontinents, Gondwana and Laurasia, lay near the equator, the climate during this period was warm, leading to growth of plants, with new types developing that could survive on dry land.

These plants did not have roots or leaves, reproduced in the absence of spores or seeds, and were only a few centimetres tall. With the evolution of vascular tissues, plants were able to carry water and food through their roots and leaves. However, the most important development in plants of the Devonian Period was the formation of seeds.

Towards the middle of this period, the first forests appeared, with trees growing upto 100 ft tall. With the appearance of vascular varieties (plants with stems and roots), plants of this period invented a new way to obtain energy. Instead of depending on the rapidly diminishing supply of organic molecules in the oceanic broth, they took in $CO_2$ and energy from the Sun via a process, which we now know as photosynthesis.

## Did You Know

A live specimen of the *coelacanth* (pronounced seel-a-canth), a lobefin fish that appeared about 350 MYA and believed to have become extinct with the dinosaurs about 65MYA, was captured in the Indian Ocean as recently as 1938.

Coelacanth

## Late Devonian extinction

By the end of this period, nearly 70 per cent of all invertebrate species were wiped out. Warm water marine species were the most affected, while terrestrial species were impacted the least. Theories suggest global cooling could be one of the causes, which may have occurred due to the re-glaciation of Gondwana and the consequent widespread lowering of the sea level and reduced atmospheric levels of $CO_2$ because of the expansion of terrestrial plants. Collision with a major asteroid has also been suggested as a probable cause of the extinction.

# CARBONIFEROUS PERIOD

The Carboniferous Period gets its name from the word "carboniferous", which means "coal bearing", as large underground coal deposits were formed during this period from vegetation. These coal deposits are found in northern Europe, mid-western and eastern North America, and Asia.

The large landmasses of Laurasia and Gondwana continued to move towards one another, resulting in a collision during the late Carboniferous Period that led to the formation of the Appalachian Mountains of North America. By the end of this period, the Earth's landmasses were gradually combining into a supercontinent called Pangaea.

Two massive ice sheets at the South Pole trapped huge amounts of water in the form of ice. Consequently, sea levels dropped and the area under land increased. Thus, in this period, terrestrial animals became more and more diverse. Vast number of tetrapods evolved during this period. Some of them were early "amphibians", some primitive reptiles with leathery skin, and an unusual group of animals called *synapsids*. These mammal-like reptiles, such as the monstrous *Dimetrodon* and *Edaphosaurus*, were the ancestors of all mammals of today.

## Time zone

- Began 359.2 MYA
- Ended 299 MYA

## What was the Earth like

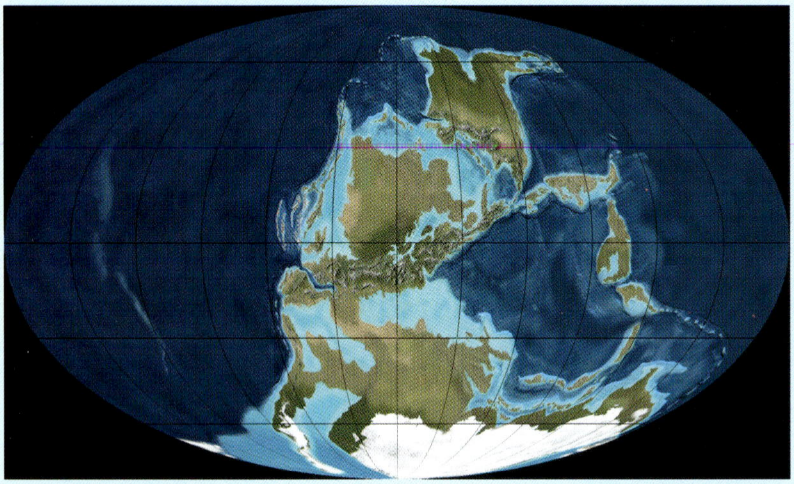

## What lived then

### CARBONIFEROUS LIFE

This period is noted for the growth in amphibians, both in size and diversity. The *Amphibamus* had a long snout and a flattened head. Occupying all types of habitat, the amphibians varied tremendously in size – from about 4 in to 20 ft long. With their skin specially adapted to not drying up in the absence of water, these amphibians were able to reduce their dependence on wetland habitats. Another important

## Did You Know

How insects first became airborne remains a mystery. One theory suggests that wings grew out of a gill-like growth present in the very first insects. An alternative theory suggests that wings arose, not from pre-existing body parts, but from special outgrowths from the body wall.

*Meganeura*

*The surplus of oxygen in the atmosphere caused creepy-crawlies to become giant-sized during the Carboniferous Period*

© Richard Bizley; www.bizleyart.com

evolutionary adaptation that helped in this regard was the amniotic egg for reproduction. These eggs had a protective covering and a fluid-retaining membrane underneath that prevented the inside from drying out as the baby developed.

**Fossil Fact**

*Protecticus ambulans*
Size: 21 mm with 14 mm wings
Fossil Site: Mazon Creek, Francis Creek Shale, Braidwood, Illinois

A prehistoric insect's Fossil

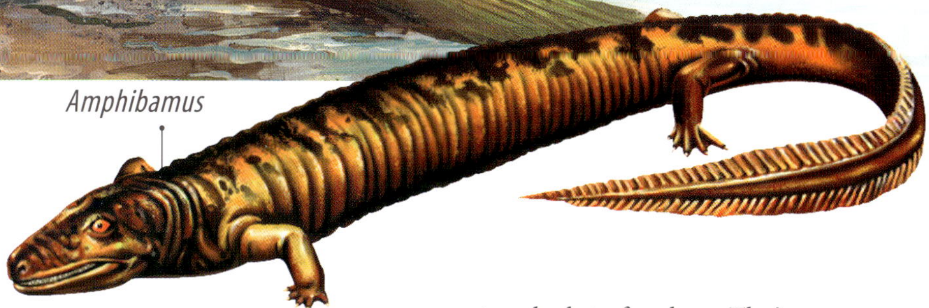

*Amphibamus*

### GIANT CREEPY-CRAWLIES

Huge cockroaches, 6-feet-long poisonous centipedes, and 3-feet-long scorpions roamed the Earth. Winged insects, such as dragonflies and mayflies, appeared for the first time. Other insects that emerged were primitive grasshoppers, crickets, and the first land scorpions.

## What **grew then**

Plants included giant club mosses, tree ferns, and great horsetails. With the growth of forests, oxygen levels soared to about 30 per cent (the current level is about 21 per cent). Lush forests grew, with trees sometimes reaching up to a height of 100 ft. Trees also captured a lot of carbon. Their dead remains, after being covered by sediment over millions of years, gave rise to the world's first extensive coal deposits, or fossils, which we are using till this day.

## Global **climate changes**

The second part of this period is marked by alternating cool and warm periods, leading to cycles of glacier expansion and retreat. Earth's climate and atmosphere during this period resemble our own climate and atmosphere like no other period and a study of this period will help us in understanding global warming and our present climate.

# PERMIAN **PERIOD**

This period is named after the Perm Oblast region of central Russia, where rocks belonging to this period were first studied. By the end of this period, all the continents of the world had more or less merged into a single supercontinent Pangaea, meaning "entire Earth". When the continent of Siberia collided into the northern part of Laurasia, the Ural Mountains were formed. While Pangaea was surrounded by the Tethys Sea, most of the remaining surface area of the planet was covered by a single ocean called Panthalassa.

Despite the presence of the ocean, Pangaea, due to its massive size, was characterized by an extreme climate. With its ice caps, the south was cold and dry, but the northern areas were intensely hot and experienced fluctuations between wet and dry conditions. Central Pangaea, where temperatures ranged from very cold to very hot, turned into deserts. The atmospheric oxygen level dropped to about 15 per cent.

## Time **zone**

- Began 299 MYA
- Ended 251 MYA

## What was the **Earth like**

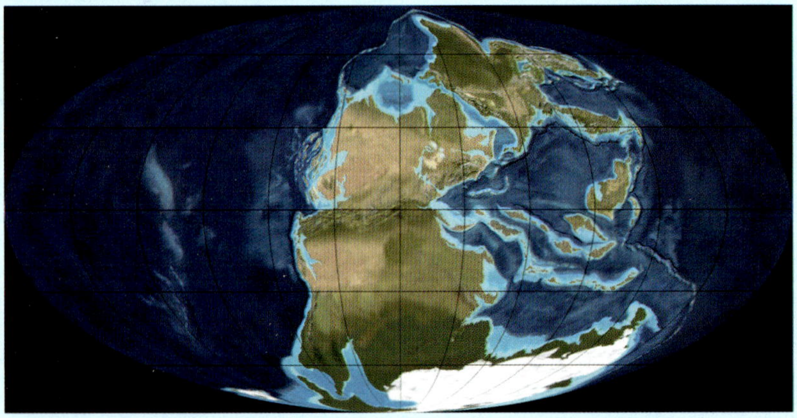

## What **lived then**

### LIFE ON LAND

Amphibians, which were dependent on moist habitat, began to die out and were replaced by reptiles – cold-blooded beings covered in scales that were better suited to the new environment due to their thick, moisture-retaining skins. Reptiles had to deal with big daily variations in temperature – below freezing at night to over 38°C during the day. Therapsids were the reptiles from which mammals evolved and became the dominant land animals towards the end of this period.

### LIFE IN WATER

The Permian seas came to be dominated by bony fish with fan-shaped fins and thick, heavy scales. There were large reef communities that harboured squid-like nautiloids. Ammonoids, with their tightly coiled spiral shells were also widespread in the Permian fossil record. The corals, which had almost become extinct at the end of the Devonian Period, began to produce huge reefs again.

### RISE OF INSECTS

With diversification in plants, insects evolved rapidly. Several new species of insects emerged during this period, such as beetles and scorpion flies – the latter being the earliest known insect to undergo metamorphosis.

*Dimetrodon*, a famous finback Pelycosaur

© Richard Bizley; www.bizleyart.com

*The changing climate affected the evolution of animals during the Permian Period*

## Fossil Fact

*Dimetrodon grandis*
Geological Time: Early to Middle Permian
Size: Up to about 10 ft
Site: North America (Texas)
*Dimetrodon,* meaning "two-measures tooth," was a reptile, more closely related to mammals than to reptiles, such as dinosaurs, lizards, and birds. *Dimetrodon* fossils have been found in North America and Europe. Growing up to 10 ft in length, and possessing a large head with large canine-like teeth, it was a top carnivore during the Permian Period.

A skeleton of *Dimetrodon grandis*

A giant dicynodont from Poland

## What grew then

As the global climate turned warm, the lush swamp forests of the last period died and were replaced by conifers, seed ferns, and other drought-resistant plants. New plant species, adapted to dry conditions, developed. These were the seed-bearing gymnosperms.

### Did You Know

*Ginkgo biloba* is one of the world's oldest living tree species, about 350 million years old! It is commonly referred to as a "living fossil".

*Ginkgo biloba* plant

## Permian mass extinction

This period ended with the worst mass extinction, which wiped out over 95 per cent of marine life and 70 per cent of land animals. Some species disappeared, while the number of sharks, bony fishes, therapsids, reptiles, and amphibians came down.

The extinction possibly resulted from a combination of two or more causes. Some scientists think volcanic ash from a series of volcanic eruptions in Siberia might have blocked the Sun, resulting in a drop in temperature and preventing plants from performing photosynthesis. Other scientists believe global climate changes or release of methane, stored under the seabed, due to earthquakes or the impact of a large extraterrestrial object like an asteroid or meteor could have caused the extinction.

*Scylacosaurus*

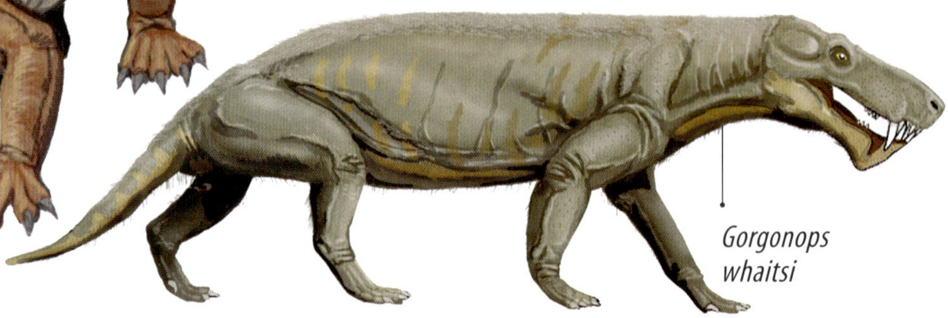

*Gorgonops whaitsi*

# GEOLOGIC TIME RANGE II

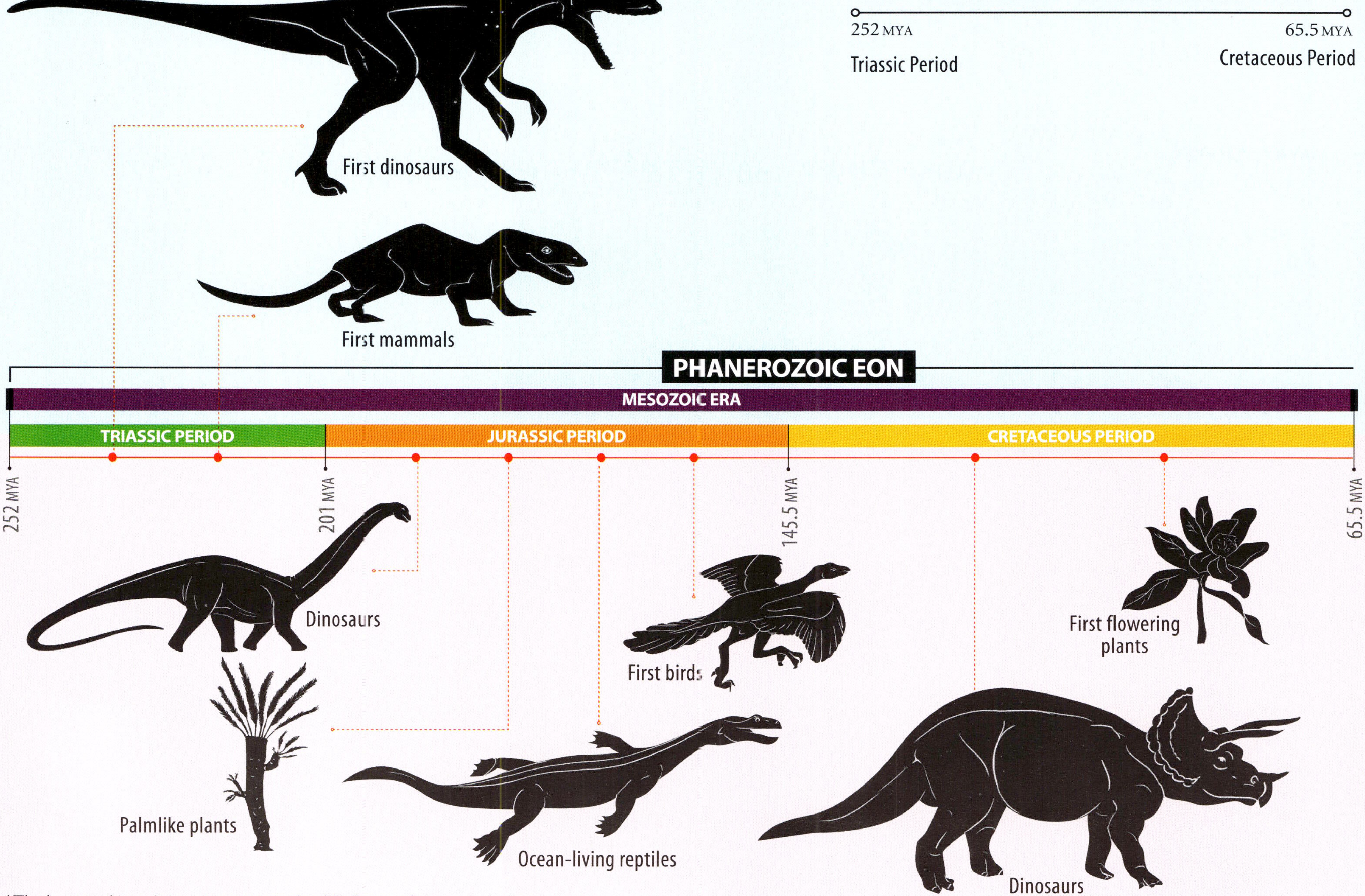


*The images shown here are representative life forms of the periods they refer to

# TRIASSIC **PERIOD**

The Triassic Period was the first of the three divisions of the Mesozoic Era. It was during this period that dinosaurs began to emerge. Three groups of animals thrived in this period – those who survived the Permian extinction, new groups that became extinct after flourishing for a brief period, and new groups that dominated the Mesozoic world – the dinosaurs belonged to the third category.

At the beginning of this period, the formation of Pangaea was complete. However, towards the end, the supercontinent had slowly begun to break apart into Laurasia (in the north) and Gondwana (in the south).

The climate was hot and dry but presence of large waterbodies, such as lakes and rivers, helped in the growth of forests that were home to large amphibians. Even the poles were covered by lush forests, because they were ice-free.

## Time **zone**

- Began 252 MYA
- Ended 201 MYA

## What was the **Earth like**

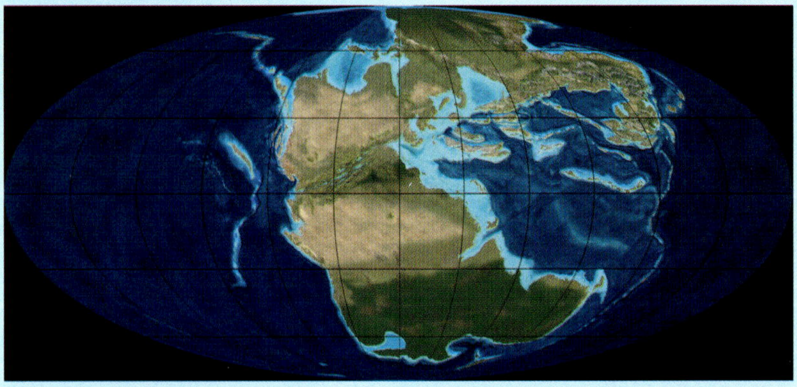

## What **lived then**

### LIFE ON LAND

The biggest change occurred with the evolution of dinosaurs and the first mammals in the Late Triassic, starting around 230 MYA. The terrestrial life was dominated by mammal-like reptiles – the pelycosaurs, archosaurs, and therapsids. The early dinosaurs found in this period were small, measuring about a metre (3.3 ft) to 4 m (13 ft). They were mainly herbivores or insectivores; so they were not in direct competition with the dinosaurs, most of whom were meat eaters and preyed on each other.

The first dinosaurs like *Herrerasaurus* (also called Herrera's lizard) and *Eoraptor* evolved along with the first prehistoric crocodiles, and pterosaurs from archosaurs.

With its distinctive sail, *Dimetrodon* was the most famous pelycosaur that ever lived.

The three-foot-long *Eozostrodon* was one of the earliest true mammals. These shrew-like creatures laid eggs, but fed their young ones with milk.

Almost 30 ft long, the *Plateosaurus* was the first truly large herbivorous dinosaur that appeared towards the end of the Triassic Period

Frogs, salamanders, crocodiles, turtles and snakes also existed. Spiders, scorpions, millipedes and centipedes prospered, while grasshoppers made their appearance.

### Did You Know

The therapsid *Cynognathus* meaning "dog-jaw" was a dinosaur that was half-tiger, half-lizard. It was a fast, fierce predator, nearly as large as a modern wolf. Unlike most of the cold-blooded reptiles of its time, evidence suggests that the *Cynognathus* were warm-blooded, had mammalian metabolism, and may have given birth to young ones.

*Cynognathus*

An artist's rendition of Triassic Period featuring a scorpionfly, a dicynodont, some gymnosperms, and small palm-like trees

*Coelophysis*

## Did You Know

Ichthyosaurs were predatory marine reptiles that evolved from an as yet unidentified land reptile that moved back into the water. They dominated the seas until they were replaced in the Cretaceous Period by the plesiosaurs. They were long-necked slow-moving marine reptiles, too slow to catch a mackerel, salmon or seal. Plesiosaurs became extinct at the same time as the dinosaurs.

Ichthyosaurs

### MARINE LIFE

The oceans teemed with the coiled-shelled ammonites, mollusks, and sea urchins and giant reptiles, such as plesiosaurs and dolphin-shaped ichthyosaurs. Microscopic plants called phytoplankton appeared for the first time. Early marine reptiles like *Placodus* and *Nothosaurus* and new species of prehistoric fish, as well as simple animals like cephalopods also appeared in the oceans.

## Fossil Fact

The fossilized bones of the oldest dinosaur were found by a goat farmer named Don Victorino Herrera and so it is named "Herrera's lizard". Its fossils were first discovered in Argentina, South America.

A skeleton of *Herrerasaurus*

### AVIAN LIFE

The only animals that flew in the skies of the early Triassic Period were insects. But towards the end of the period, the earliest pterosaurs, such as the *Eudimorphodon*, took to the air and probably just hopped around the shores of Europe.

## Triassic mass extinction

The causes of the Triassic extinction are not well known, but popular explanations for its occurrence include global climatic cooling, comet showers, and volcanic activity.

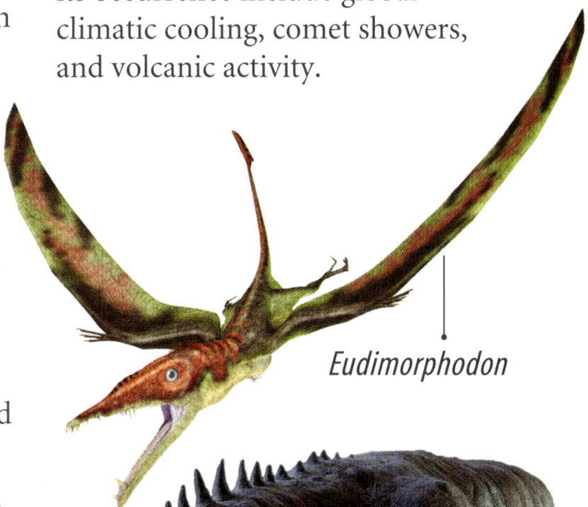

*Eudimorphodon*

*Herrerasaurus*

## What grew then

This period was not as lush and green as the later Mesozoic Period, but it did see an explosion of various land-dwelling plants, including cycads, ferns, and Ginkgo-like trees.

# JURASSIC **PERIOD**

The Jurassic Period was the second of the three divisions of the Mesozoic Era. This period was known for the most diverse range of dinosaurs the world has ever seen.

The disintegration of Pangaea sped up during this period. Continents were covered by shallow seas, and large lakes and rivers provided new homes for various aquatic organisms and made the climate humid. The vegetation-covered land offered an ideal environment for gigantic terrestrial animals to flourish. Average global temperatures were as much as 10 degrees higher than now and there was steady rainfall that led to an explosion of lush, green plants.

## Time **zone**

- Began 201 MYA
- Ended 145.5 MYA

## What was the **Earth like**

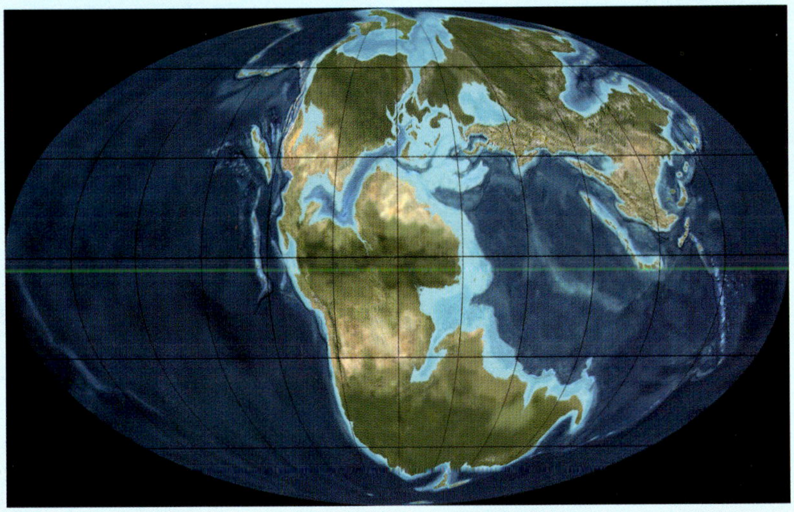

## What **lived then**

### LIFE ON LAND

For a period of 185 million years, dinosaurs were rulers of the animal kingdom, in water and on land.

Gigantic herbivorous groups of dinosaurs called sauropods were relatives of the small, four-legged, plant-eating prosauropods of the Triassic Period. Sauropods such as *Apatosaurus* had massive elephant-like bodies, long necks and tails, and walked on all fours and survived for around 100 million years. The largest sauropods weighed almost 10 times the weight of a modern elephant, and were amongst the largest land animals of all time. *Diplodocus* measured 90 ft and weighed about 15 tonnes.

The *Argentinosaurus* was the largest, heaviest, and longest land animal of all time. It stood 70 ft tall, measured up to 115 ft in length, and weighed up to 100 tonnes.

This period also saw the rise of a group of predator dinosaurs called theropods, meaning "beast footed", which included all flesh-eating dinosaurs like the *Allosaurus*, *Megalosaurus*, and *Tyrannosaurus* among others. Today, birds are the only living descendants of theropods. For over 100 million years, they were the only large terrestrial carnivores. However, with the passage of time, some of these carnivorous creatures adapted themselves to an omnivorous or herbivorous diet.

The *Allosaurus* was a large carnivorous dinosaur that weighed 2 tonnes and grew up to

*Ceratosaurus*

40 ft in length. It had an enormous jaw with dozens of long, sharp, saw-like teeth, a tail that was as long as half its body, and two legs. It preyed on large herbivorous dinosaurs as well as other predators, including its biggest rival *Ceratosaurus*.

The *Ceratosaurus* had a blade-like horn on its snout and long dagger-like teeth. Its powerful legs and feet helped it escape from predators and overpower its prey.

To protect itself, the *Ankylosaurus* used its thick armour made up of flat, bony plates along its body. Additionally, it had a long tail that ended in a club-shaped bone, which was probably swung as defence against predators.

*Stegosaurus* was instantly recognizable due to its spiked tail. Sometimes growing up to 30 ft in length, its tail ended in two pairs of pointed bony spikes that most probably acted as defensive weapons.

Other reptiles like turtles, were also present, while modern forms of lizards appeared towards the end of the period. Insects like leafhoppers and beetles were present in large numbers, while some of the world's earliest mammals walked the Earth.

## AVIAN LIFE

Though the debate on the origin of birds has not reached a definitive conclusion, scientists have reached consensus on the

© Richard Bizley; www.bizleyart.com

*At the lower left is Ardeosaurus, an early gecko. In the sky is Pterodactylus, a flying reptile. On the land are four dinosaurs, from left to right: Stegosaurus, Brachiosaurus, Apatosaurus, and Allosaurus.*

fact that the earliest ancestors of birds were small theropods from the Jurassic Period.

*Epidendrosaurus* was the smallest of all dinosaurs, known to be the first reptile that was more a bird than a dinosaur. It lived on trees and hopped from branch to branch.

*Archaeopteryx* had distinctly reptilian characteristics like a lizard-like tail, teeth, and three claws jutting out from the middle of each wing.

*Pterodactyl* was sparrow-sized with short tails. Dimorphodons had huge heads and two distinct types of teeth. They had light skeletons, and were specially adapted to flying and gliding.

Liopleurodon

## MARINE LIFE

The seas were filled with fierce marine reptiles with flippers called plesiosaurs.

*Plesiosaurus*, an early plesiosaur, grew up to 15 ft long and had a flexible neck and sharp teeth to capture its prey.

*Liopleurodon*, the mightiest marine predator of them all, sometimes growing to over 60 ft in length, was an expert hunter, known for its ability to smell its prey under water.

The oceans were also teeming with massive crocodiles, prehistoric sharks and rays, fish-like ichthyosaurs, squid-like cephalopods, and ammonites with coiled shells. This period is also noted for the appearance of a sub-category of bony fish, called teleost. Most living fish species today belong to this sub-category.

## What grew then

The land was covered with thick vegetation, including ferns, conifers, cycads, club mosses, and horsetails. Forests grew even near the poles.

### Fossil Fact

The first dinosaur fossil to be discovered was of *Megalosaurus* or "great lizard" in England in 1676. They were large carnivorous predators up to 30 ft long (9 m), 10 ft tall (3 m), and weighed about one tonne. They had two powerful legs, a strong, short neck, a big head with sharp, saw-like teeth and a bulky body that ended in a massive tail. Its short arms had three-fingered hands with sharp claws.

A skeleton of Megalosaurus

## Jurassic mass extinction

Two extinction events are speculated to have occurred in the Jurassic. The first extinction eliminated more than eighty per cent of marine and shallow water species. The second crisis occurred near the end of the Jurassic that not only affected marine life but dinosaurs as well.

*Ankylosaurus*

# CRETACEOUS **PERIOD**

The Cretaceous Period is the last of the three periods that constitute the Mesozoic Era. The word Cretaceous comes from the Latin word "creta", which means chalk. It was the last age of the dinosaurs, as these fascinating creatures were to become extinct at the end of that time.

## Time **zone**

- Began 145.5 MYA
- Ended 65.5 MYA

## What was the **Earth like**

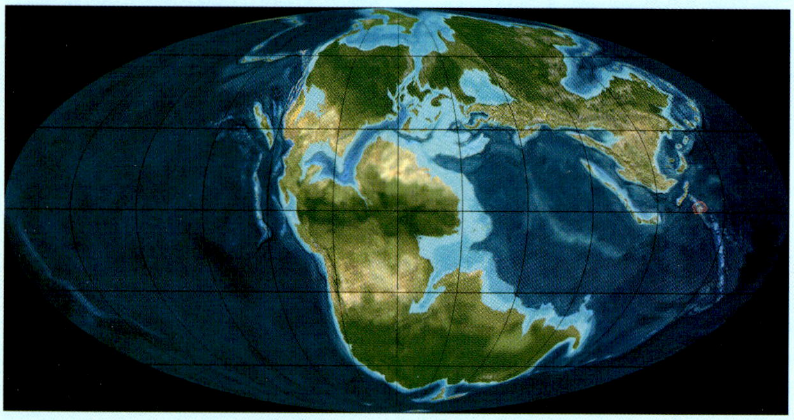

### GEOLOGICAL FEATURES

The continental landmasses underwent several significant changes during this period that lasted for 80 million years.

The second major episode of continental rifting began with the breakup of Pangaea during the Late Jurassic Period, and continued into the Cretaceous. South America and Africa separated slowly from south to north, creating the South Atlantic Ocean, while India and Madagascar rifted away from the western margins of Australia and Antarctica to form the Indian Ocean.

As Laurasia and Gondwana began to drift apart, the North Atlantic Ocean was formed towards the east of Laurasia. On the other hand, South America and Africa split from Gondwana. A rift led to the formation of the South Atlantic Ocean, while another rift towards the east of Africa resulted in a water body between Africa and another landmass that comprises present-day Madagascar and India. With North and South America splitting from Europe and Africa, the Atlantic Ocean became wider. It was during the Cretaceous Period that the Indian Ocean was formed.

The splitting of Africa from South America led to the formation of a number of underwater mountain ranges. With the expansion of the Atlantic Ocean, these mountains displaced huge amounts of water, causing the sea level to rise higher than any other period in the history of our planet. It is estimated that during the Early Cretaceous Period, the oceans were about 330–660 ft higher than their present state, and about 660–820 ft higher during the Late Cretaceous Period than the current levels. As compared to 28 per cent today, only about 18 per cent of the Earth's surface was covered by land.

Generally, the climate was a lot warmer than what it is now, but rising sea levels and the spread of swamps allowed dinosaurs and other prehistoric life to prosper.

## What **lived then**

### LIFE ON LAND

Dinosaurs continued to rule the land. The innumerable varieties of these creatures that roamed the Earth included the theropods. The ostrich-like ornithomimids meaning "bird mimics" were fast-running creatures with uniquely structured hands, long slim hind limbs, flexible neck, and a small beak. The strange *Therizinosaurus*, a herbivorous creature with a long neck, protruding belly, small tail, four-toed feet, a beak, gigantic claws on its forelimbs, and possibly, feathers, was a combination of about half a dozen other dinosaurs.

Then there was one of the smartest dinosaurs, *Troodon*, meaning "wounding tooth", which gets its name from its sharp, curved, saw-like teeth. A small carnivore about 6.5 ft long, it weighed about 50 kg, with a large brain for its size. It walked on two hind legs, had three-fingers on each of its hands, and was known for its sharp vision.

*Raptors* were small- to medium-sized bird-like, carnivorous dinosaurs with distinctive, large, single claws on each of their hind feet that came to the fore during the Late Cretaceous Period. *Utahraptor* was a vicious creature, which used its 9.5-inch-long sickle-shaped claws to rip its prey apart.

*Ornithischians*, meaning "bird-hipped", were a group of herbivorous dinosaurs, related closer to birds than to dinosaurs and found everywhere except in Antarctica. They lived in herds and were toothless with horny beak and powerful teeth, which helped them munch on plants. *Iguanodon* were bulky herbivores, had iguana-like teeth and were 30 ft long with unusual five-fingered hands. Duck-billed dinosaurs, called *Hadrosaurs*, had heads resembling

## Did You Know

The Magnolias we see today are prehistoric plants. Since they appeared before bees, magnolias slowly changed their flowers to become more attractive to bees, to ensure successful pollination. They flower in white, yellow, pinkish, purple and green colours, depending on the species.

*Sweetbay Magnolia*

*The Cretaceous Period was also the era when life on Earth, as we know it now, was formed*

© Richard Bizley; www.bizleyart.com

those of modern ducks, and were especially common in Europe, Asia, and North America.

Towards the end of the Cretaceous Period, large groups of horned beasts such as the herbivorous *Triceratops*, or "three-horned face" in Latin, lived in the northern continents. The *Triceratops* was a large four-footed dinosaur that had a bony frill at the back of its extraordinarily large head, nearly 10 ft long, and three distinctive horns placed above each eye and on the snout. The horns and frill changed shape as the creature got older, and it had columns of teeth in its upper and lower jaws and a beak that might have been used to slice vegetation. Weighing a massive 5,450 –7,260 kg, they were one of the last groups of non-avian dinosaurs to evolve. The *Triceratops* is often depicted as using its horns to defend itself from meat-eating dinosaurs like the *Tyrannosaurs*.

The *Tyrannosaurs* were the most challenging predators with a large skull that could be up to 4 ft long. The most famous and biggest *Tyrannosaur* was the *T. rex*, which walked on two legs, stood 21 ft tall, was 46 ft long, and weighed between 4,000 –7,000 kg. The *T. rex* had a huge mouth lined with about 60 teeth, each about 6 in long that could easily rip through the skulls and limbs of other dinosaurs.

Alongside dinosaurs, many new types of animals were evolving. Frogs, salamanders, turtles, crocodiles, and snakes were found in abundance on the expanded coasts, while mouse-like mammals scampered around in the forests. Two important groups of modern mammals appeared during the Cretaceous Period. These were the placental mammals which include rodents, whales and primates; and marsupials such as kangaroos, koalas, and opossum.

## Fossil Fact

The first *Tyrannosaurus rex* fossil was discovered by the famous fossil hunter Barnum Brown in 1902. About 30 fossils of different *Tyrannosaurus* species have been found in the USA.

A skeleton of *T. rex*

## AVIAN LIFE

The pterosaurs dominated the skies and by the end of this period, they had become as large as the creatures inhabiting the land and sea. The giant *Quetzalcoatlus*, a pterosaur, with a wingspan of about 49 ft, was the largest creature ever to take to the skies. However, by the end of this period, pterosaurs were getting pushed out of the sky by prehistoric birds. Originating from terrestrial feathered dinosaurs, these fast diversifying birds were the ancestors of today's grebes, cormorants, pelicans, and sandpipers.

## MARINE LIFE

Ichthyosaurs of the Triassic and Jurassic Periods had all but vanished from the oceans and the plesiosaurs were replaced by the gigantic snake-like mosasaurs, related to today's monitor lizards, They grew up to 55 ft.

Rays, sharks, sea urchins, and starfish, and bony, ray-finned fish, the teleost, were abundant. One specific type of plankton called the coccolithophores became

**Fossil Fact**

The monstrous Spinosaurus. was probably the largest carnivore to walk the Earth. It was known for its 5-feet-long spine that ran along its back and formed its most fascinating feature, the sail-like fin. Its long, narrow skull is quite similar to that of today's crocodiles; and just like crocodiles, the Spinosaurus lived and hunted both in water and on land. However, its original fossils were destroyed in World War II, but those discovered by Ernst Stromer in Egypt shortly before World War I were sent to Deutsches Museum in Munich where they were destroyed by an Allied bombing raid in 1944.

A skeleton of Spinosaurus

so commonplace that their vast quantities formed the substance from which the Cretaceous Period derives its name – chalk. A new group of marine organisms called diatoms that were capable of photosynthesis made their appearance during this era.

## What grew then

Rich vegetation from the earlier period continued to flourish with conifers and cycads. However, this period marked the arrival of angiosperms or flowering plants like magnolia, ficus, and sassafras that led to the appearance of the oldest-known butterflies. Aphids, grasshoppers, and wasps, termites, and ants were other insects that were also seen for the first time.

## Cretaceous mass extinction

### EXTINCTION OF DINOSAURS

The most well known of all mass extinctions, in which about 80 per cent of all animals (except birds), including the dinosaurs, disappeared from the face of the Earth, brought the Cretaceous Period to a close. This event is known as the Cretaceous–Tertiary event or the K–T event, where "K" stands for "Cretaceous". Many theories have been put forward to explain the K–T event, some of which are mentioned below:

- The great retreat of the seas led to the drying of fertile swamps, reducing food-rich zones.
- Massive outburst of gases from the Sun produced intense radiation that caused glandular disorders among dinosaurs and also damaged the flora.
- Poisonous alkaloids may have been present in flowering plants that killed the plant-eating dinosaurs.
- A giant asteroid struck the Yucatan peninsula in Mexico, and this might have thrown up a vast amount of rock debris into the atmosphere, covering the Earth in darkness for several months and crippling the process of photosynthesis.
- Massive outpouring of lava during the later Cretaceous Period released huge amounts of $CO_2$ into the atmosphere. This warmed the planet to an extent unbearable for the organisms to survive.
- Tectonic movements and the breakup of Pangaea led to climate change.

Although all land animals that weighed over 25 kg became extinct, many kinds of animals, including fish, frogs, turtles, crocodiles, birds, and mammals survived and flourished over the next few million years.

# GEOLOGIC TIME RANGE III

65.5 MYA — Continues until now

Paleogene Period — Quaternary Period

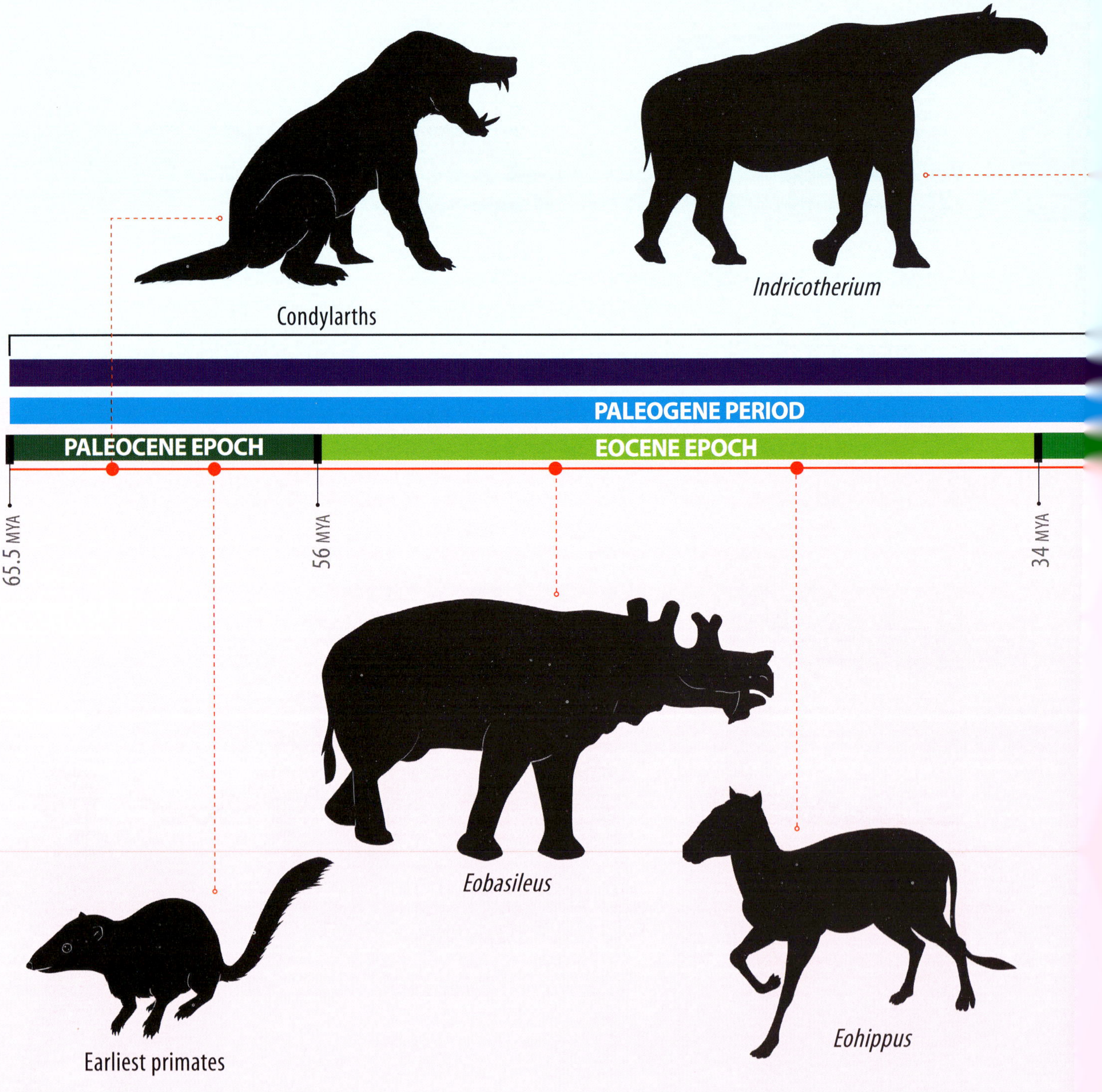

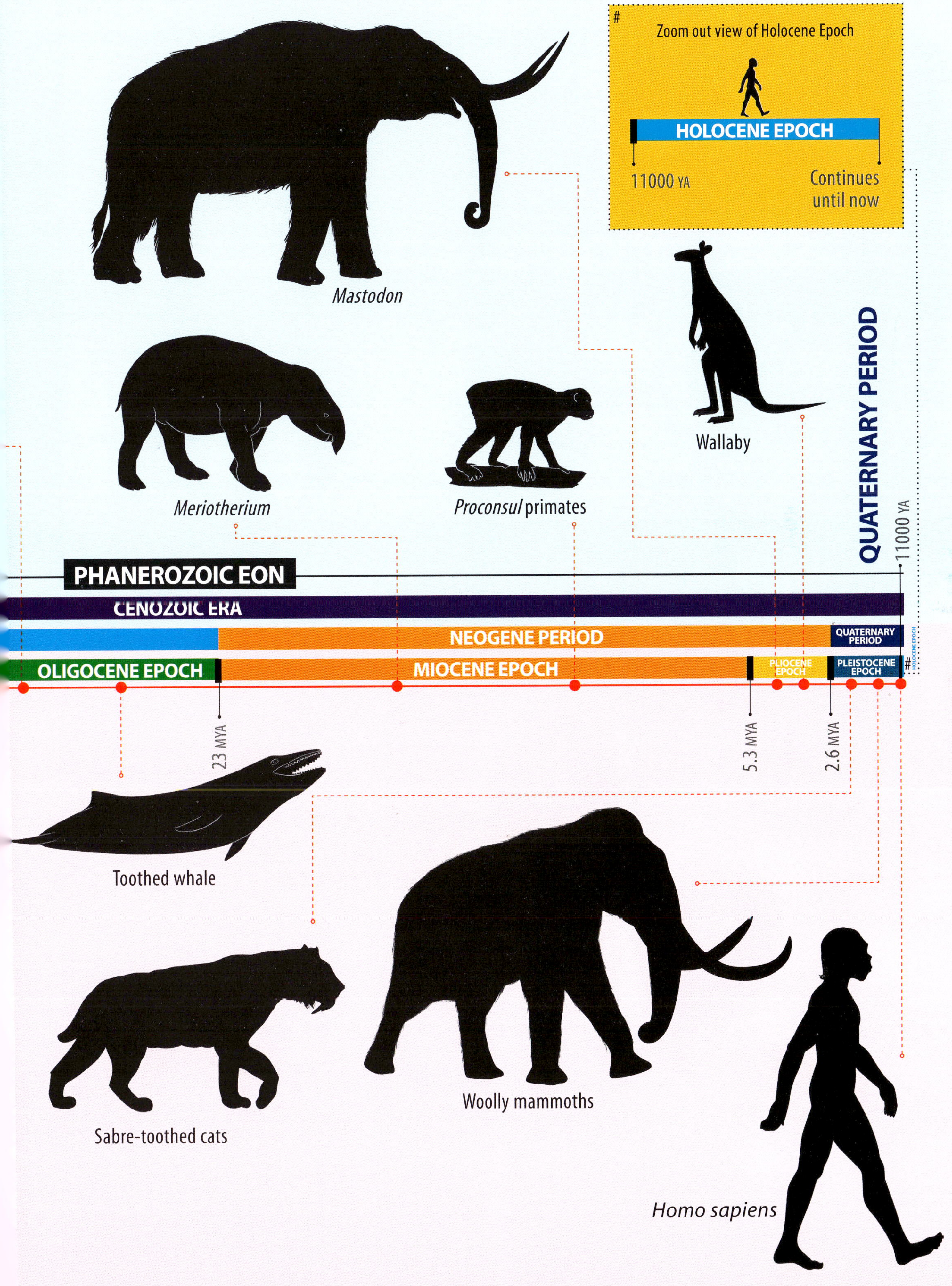

#
Zoom out view of Holocene Epoch
HOLOCENE EPOCH
11000 YA
Continues until now
Mastodon
Meriotherium
Proconsul primates
Wallaby
QUATERNARY PERIOD
11000 YA
PHANEROZOIC EON
CENOZOIC ERA
NEOGENE PERIOD
QUATERNARY PERIOD
OLIGOCENE EPOCH
MIOCENE EPOCH
PLIOCENE EPOCH
PLEISTOCENE EPOCH
#
23 MYA
5.3 MYA
2.6 MYA
Toothed whale
Sabre-toothed cats
Woolly mammoths
Homo sapiens

# PALEOGENE PERIOD

Paleogene means "ancient born" in Greek. It is the oldest of the three divisions of the Cenozoic Era. Dinosaurs, pterosaurs, and giant marine reptiles had become extinct by the beginning of this period. With their disappearance, the nocturnal rodent-like mammals, which had, till then, scurried into the forests at night, found it safe to step out into the open. Over the next 42.5 million years or so, these mammals began to grow in size, number, and variety. And as this period drew to a close, life forms found even today filled the seas, land, and air.

## Time zone

- Began 65.5 MYA
- Ended 23 MYA

## What was the Earth like

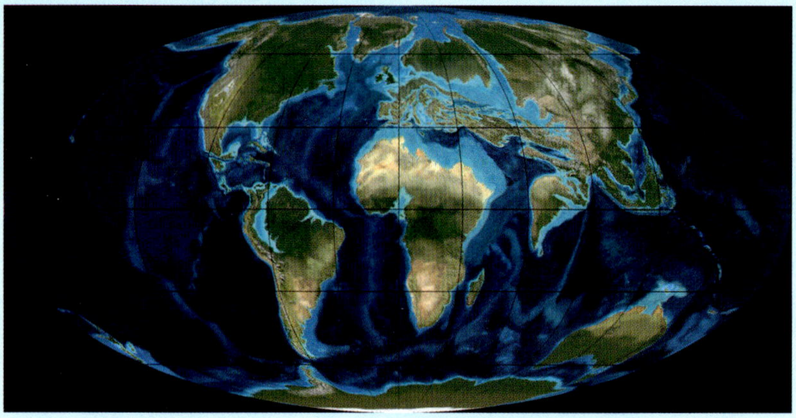

### CONTINENTS

During this period, the continents slowly drifted towards their present positions. Towards the late Paleocene Epoch, India collided with southern Asia. Australia separated from Antarctica and moved away. A seaway that connected the Atlantic with the Pacific and Indian Ocean separated North America from South America. South America was joined to Antarctica by a narrow peninsula. The Atlantic Ocean continued to expand, while the Pacific Ocean shrank. North America was connected to Asia, Europe, and Greenland. Between Europe and Africa lay the Tethys Sea. The origin of the Alps, Carpathian Mountains, and Atlas mountains in Africa and the formation of glaciers on Antarctica around 35 MYA can be traced back to this period.

### CLIMATE

During the beginning of the Paleogene Period, the global climate was a lot warmer than what it is today. Towards the end of the Paleocene and beginning of the Eocene Epoch, a huge volume of methane stored in sea floor sediments released into the atmosphere. Since methane is a greenhouse gas, its increased concentration in the atmosphere implied higher amounts of the Sun's heat getting trapped. As methane reached the atmosphere, it combined with oxygen to form $CO_2$ – another potent greenhouse gas. The result was an intense period of "global warming", when sea-surface temperature rose between 5°C and 8°C in just a few thousand years. This abrupt change in climatic conditions is known as the Paleocene-Eocene Thermal Maximum (PETM). It killed off 50 per cent of unicellular marine organisms called *foraminifera*, alongside other invertebrates.

This period of warming that reached its peak about 55 MYA was followed by gradual cooling and a dry climate.

### LIFE

Animals that had survived the K-T extinction, such as turtles, snakes, crocodiles, and lizards abounded the coastlines. However, the major characteristic feature of this period was the rapid evolution of mammals.

Birds, not mammals, were the dominant animals during the early part of the Paleogene Period. During the Paleocene and Eocene epochs, mammals were still fairly small, but they had already started evolving. By the end of the Paleogene, primates, horses, bats, pigs, dogs and cats had all evolved. Flowering plants and deciduous forests flourished, and the first grasses made their appearance towards the end of this period.

© Richard Bizley; www.bizleyart.com

*This period marked a crucial interval in the evolution of mammals, birds, and reptiles, after the end of dinosaurs in the Mesozoic Era*

# PALEOCENE **EPOCH**

Paleocene means "old recent" in Greek. At the beginning of the Paleocene Epoch – the first subdivision of the Paleogene Period – the Earth was still recovering from the impacts of the K-T extinction event. The climate was subtropical and oceans were warm, while there were no ice caps at the poles. By the end of this epoch, rapid warming occurred that led to the extinction and reorganization of various terrestrial and marine species. North America, Asia and Europe were connected to one another via land bridges. South America was connected to Antarctica, Africa, Australia, and India were all island continents or subcontinents.

### Time **zone**

- Began 65 MYA
- Ended 56 MYA

### Did You Know

A range of now-extinct mammals, Condylarths, varied in size from 1–4 ft in length and are believed to be the ancestors of many modern mammals. They gave birth to live young and had carnivore-like teeth, but scientists believe they were omnivores. They had clawed feet and were the first animals to evolve hoofed edges, which helped them walk on rough terrain.

Condylarth

## What **lived then**

### RISE OF THE MAMMALS

Although mammals had appeared as early as the Triassic Period, and developed alongside the dinosaurs, they lived in the insect-rich forest undergrowth or high up in the trees. Early mammals were small and fed on plants and insects. They were generally short-legged, yet to develop specialized teeth or limbs, had five toes on each foot, and walked on the soles of their feet. The largest mammal was the herbivorous *Pantolambda*. Hoofed condylarths and multituberculates, the herbivorous mammals with rodent-like teeth, were common. Creodonts were carnivorous mammals that resembled cats and dogs. Primates became quite abundant in the middle of the epoch. *Phosphatherium* was the earliest known ancestor of elephants. A dog-sized herbivore, it weighed about 45 kg.

Egg-laying mammals like the Platypus, and two species of Echidnas, or spiny ant eaters, also emerged.

Echidna

### REPTILES

Turtles were the most abundant and diverse of all reptiles. They flourished particularly in freshwater streams and lakes.

Crocodiles were the largest beings to survive the K-T extinction event. Lizards and snakes, especially those that fed on small rodents, continued to diversify. The largest snake ever, the *Titanoboa*, was over 40 ft long, and weighed over a ton.

### AVIAN LIFE

Ancestors of most modern birds such as cranes, hawks, owls, ducks, pigeons, penguins, and woodpeckers appeared. Fossil remains suggest the existence of large carnivorous flightless birds, popularly known as terror birds, that dominated South America from about 65–2.5 MYA. The frightening *Gastornis*, formerly known as Diatryma, was a massive creature, standing

## Did You Know

Purgatorius is the world's earliest and most primitive primate that was squirrel-like in size and appearance. This tree-dwelling primate arrived on the scene during the early Paleocene Epoch.

*Purgatorius*

about 6.5 ft tall. It was armed with a huge beak and formidable talons on its toes. Other terror birds included the phorusrhacids. The smallest known terror bird weighed about 8 kg, while the largest one was almost 10 ft tall and weighed about 500 kg!

## Fossil Fact

*Plesiadapis* was a lemur-like primate about 2.5 ft long with a long tail and eyes on the sides of the head. An important difference with modern primates was the presence of a long snout with rodent-like jaws and teeth.

A fossil of *Plesiadapis*, Brussels

## PRIMATES

Fossils of the earliest primates belong to the Late Paleocene Epoch. Primates arose from ancestors that lived in the trees of tropical forests. They include 300 or more species of lemurs, lorises, tarsiers, monkeys, apes, and humans. There are two basic divisions of primates: the *Strepsirhini* and the *Haplorhini*.

*STREPSIRHINI* are known as "wet-nosed primates" as they have a moist surface around their nostrils.

**Lemurs** evolved in near isolation after Madagascar split away from the African mainland. They cannot hang by their tails from trees like monkeys. The smallest primate is Madame Berthe's mouse lemur of Madagascar that weighs about 35gm.

**Galagos** or Bushbabies are almost squirrel-sized primates. These nocturnal, arboreal primates have large eyes and foldable ears, long hind legs, soft, woolly fur, and long tails.

**Pottos** are nocturnal, tree-dwelling, slow-moving primates. Their unique feature is a spiny "shield" on the neck.

*HAPLORHINI* are commonly referred to as "dry-nosed primates". Humans too belong to this group.

**Tarsiers** are small, chiefly nocturnal and tree-dwelling primates. Of all mammals, they have the longest hind limbs relative to body length.

**Apes** include gibbons, siamangs, gorillas, chimpanzees, orangutans, and humans. Apes differ from all other primates in that they do not have external tails. They also are more intelligent and more dependent for survival on learned behaviour patterns. The largest primate is the gorilla, which can weigh up to 180 kg.

## Did You Know

Currently, the Earth is warming about 10 times faster! If $CO_2$ emissions from fossil fuels continue at the present rate, then by the year 2400, about five trillion metric tonnes of carbon would be added to the atmosphere. That is roughly the same amount of carbon that led to the PETM. However, there is one important difference. What happened in 170,000 years during the Paleogene Period will now take less than 300 more years!

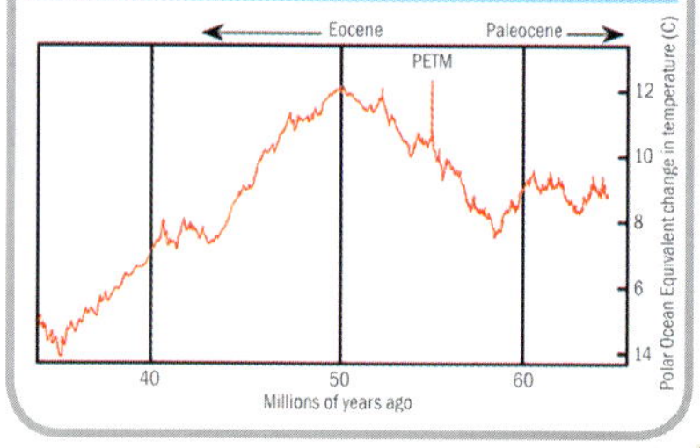

*Paraphysornis brasiliensis*

### MARINE LIFE

By the end of the Cretaceous Period, the giant mosasaurs, plesiosaurs, and pliosaurs had disappeared from the world's oceans. In their place, prehistoric sharks became the topmost predator, feeding on fish that were abundantly available. Squids, sea snails, sea urchins, and new species of shell-like *Foraminifera* appeared. But the most significant development in the oceans was the appearance of whales, which evolved from terrestrial mammals that took to the sea.

## What grew then

The climate changes following the K-T extinction event destroyed the majority of terrestrial and aquatic plants. For a brief period after the end of the Cretaceous period, ferns flourished. Then, during the early Paleocene Epoch, flowering plants first seen in the Cretaceous Period became abundant alongside conifers. Modern plant species like cacti, palm trees, and ferns also appeared. The warm temperatures nurtured a thick tropical, sub-tropical and deciduous forest cover across the globe. Even the ice-free polar regions were covered with coniferous and deciduous trees.

## Paleocene-eocene thermal maximum

The Paleocene-Eocene Thermal Maximum (PETM), which occurred around 56 MYA, is the most recent event that can be compared to the global warming the Earth is experiencing today. The PETM lasted for about 170,000 years during which the effect on the Earth's climate was so severe that a new geological era was born – the Eocene Epoch.

## Fossil Fact

*Altiatlasius koulchii* fossil was found in Morocco. It evolved during the late Paleocene Epoch. It was squirrel-like in size and appearance, but had grasping hands and feet that helped it manipulate objects and climb trees.

*Altiatlasius koulchii*

# EOCENE EPOCH

The Eocene means "dawn of the recent", which is derived from the Greek words *eos* (dawn) and *kainos* (recent). It refers to the appearance and diversification of various organisms, such as mammals. The world's continents had almost moved into their present position and giant mountain ranges continued to form. India continued to move northward and by the late Eocene, its collision with the Eurasian continent gave rise to the Himalayas.

## Time zone

- Began 56 MYA
- Ended 34 MYA

## What was the Earth like

The onset of the Eocene Epoch was a time of global warming brought about by the release of methane that had been trapped in sediments on the ocean floor. In fact, the first 5 million years of this epoch are believed to have the highest average annual temperatures of the entire Cenozoic Era. In the absence of ice caps, forests thrived and trees came up even in the Polar regions. However, creation of a deep water passage between Antarctica and Australia led to eventual global cooling in the later part of the epoch. This resulted in the re-formation of ice caps at both the poles during the succeeding Oligocene Epoch.

The ice caps of the late Eocene expanded during the Oligocene Epoch, resulting in a marked cooling across the globe and a fall in sea level. In general, the climate of this epoch was temperate, with several regions enjoying subtropical climatic conditions.

### Fossil Fact

The first fossils of this tiny horse also known as "dawn horse" was found in England by the famous paleontologist Richard Owen in 1841 and he named it *Hyracotherium*.

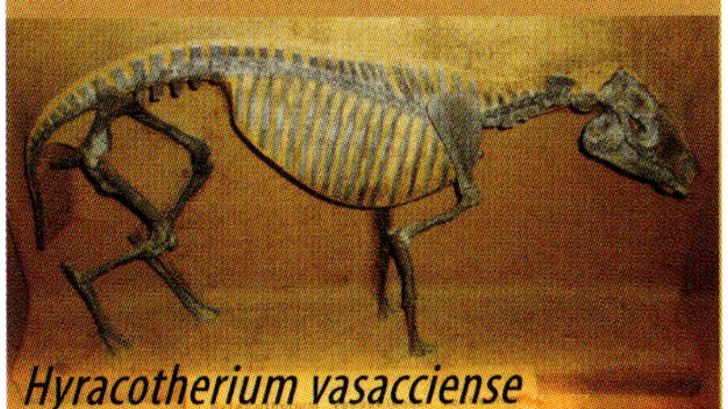

*Hyracotherium vasacciense*

## What lived then

### TERRESTRIAL LIFE

This epoch saw the appearance of the first ancestors and close relatives of modern mammal species, including horses, deer, camels, elephants, cats, dogs, and primates. The beginning of the Eocene is characterized by the appearance of two new groups of mammals, namely perissodactyls – odd-toed hoofed mammals such as horses, rhinos, and tapirs – and artiodactyls – even-toed hoofed mammals comprising deer, cattle, sheep, rodents, and primates. The ancestors of elephants known as proboscideans also belong to this group.

At the start of the Eocene, the oldest known horse, *Hyracotherium*, or the *Eohippus* appeared. It resembled a dog in size, had four hoofed toes on its front legs and three hoofed toes on its hind legs. This herbivore fed on fruits, soft leaves, and plant shoots, and roamed around like a modern deer.

### Did You Know

*Uintatherium* or "Beast of the Uninta Mountains" was an extraordinary creature of this epoch. This hoofed mammal was as big as a modern rhino and had strong limbs to support its massive body. There were six bony growths on its skull. Male uintatheriums had large, powerful canine teeth.

*Uintatherium anceps*

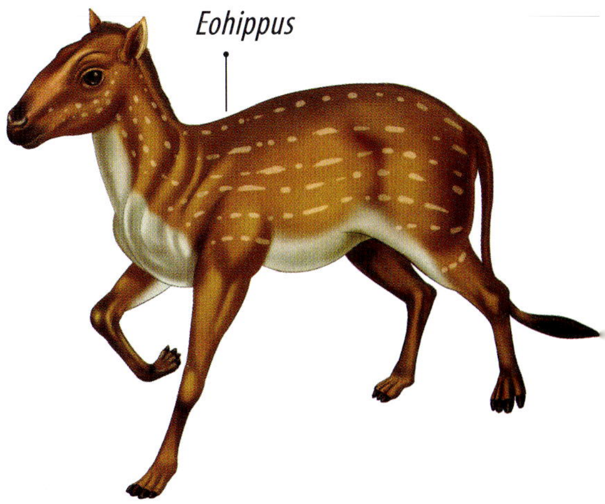

*Eohippus*

An artist's rendition of Eocene fauna in North America

The largest mammals of the epoch were the titanotheres, also known as brontotheres. They were rhino-like, herbivorous, hoofed mammals, had a heavy body supported by thick legs that looked like pillars. They became extinct towards the middle of the Oligocene Epoch. Another example of brontotheres was the *Embolotherium*, meaning "battering ram beast".

Along with these herbivores came carnivorous predators like the *Andrewsarchus*. Measuring over 16 ft from nose to tail, it was the largest-ever meat-eating terrestrial mammal.

The first recognizable bats also evolved during this epoch. The 14-cm-long insectivorous *Icaronycteris* was very similar to modern bats.

Presbyornis

Anthropornis

## AVIAN LIFE

Like mammals, the roots of many modern orders of birds can also be traced to this epoch. Modern birds like eagles, pelicans, quails, and vultures appeared. Giant ancient penguins, such as the *Inkayacu* of South America with reddish-brown and grey feathers and the *Anthropornis* of Australia that weighed up to 90 kg lived in temperate climates. Another curious bird of the era was the *Presbyornis*, a prehistoric duck with a long neck and legs and a flat bill.

## MARINE LIFE

This epoch saw the emergence of two marine mammal groups. These were the cetaceans, such as whales, porpoises, and dolphins, and sirenians that resembled present-day manatees and dugongs.

The earliest ancestors of modern whales, the *Pakicetus* had ears that were more suited to hearing in the air, so it spent more time on land than in water. The gigantic, carnivorous whales *Basilosaurus* or "king lizards" grew up to 65 ft in length. These were mistaken to be sea serpents.

Sea cows appeared in the oceans and sharks continued to evolve while bony fish populated the seas.

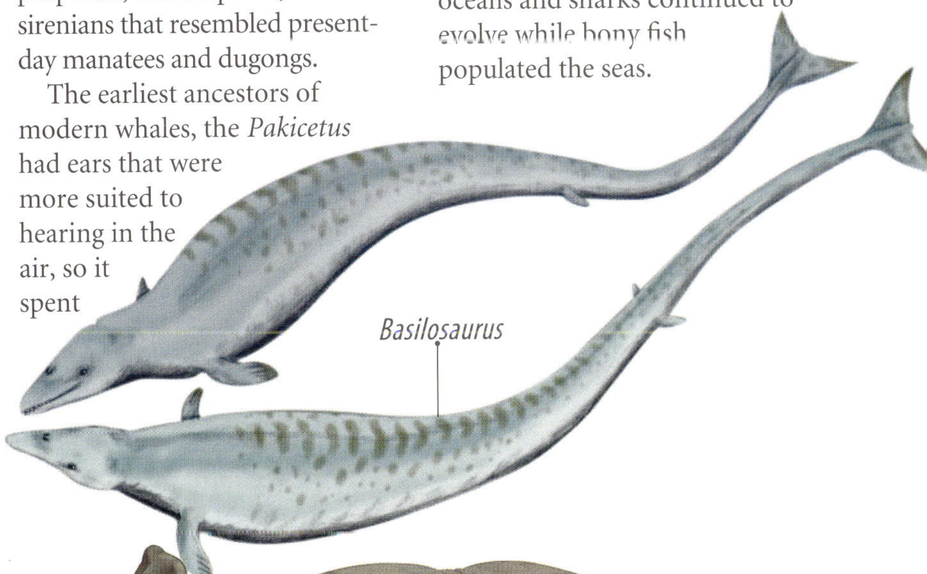

Basilosaurus

### Did You Know

Related to the artiodactyls, the prehistoric whales depict the transition from a terrestrial to an aquatic life. The 12 ft-long Ambulocetus was a "walking whale" that used its strong forelimbs to pull its body on land.

Ambulocetus

### Fossil Fact

The most important fossil find of the epoch was that of an immature female primate *Darwinius masillae*, nicknamed Ida. This 47-million-year-old, about 2-feet-long (including tail) fossil was found in a well-preserved condition in Germany.

Darwinius masillae

Icaronycteris

Embolotherium

# OLIGOCENE **EPOCH**

The Oligocene Epoch followed the Eocene Epoch. Oligocene means "epoch of few recent life forms". This epoch was characterized by an increasing proportion of "modern" groups of carnivores and herbivores. And it saw an increase in volcanic activity, especially in the Rocky Mountain region.

### Time **zone**

- Began 34 MYA
- Ended 23 MYA

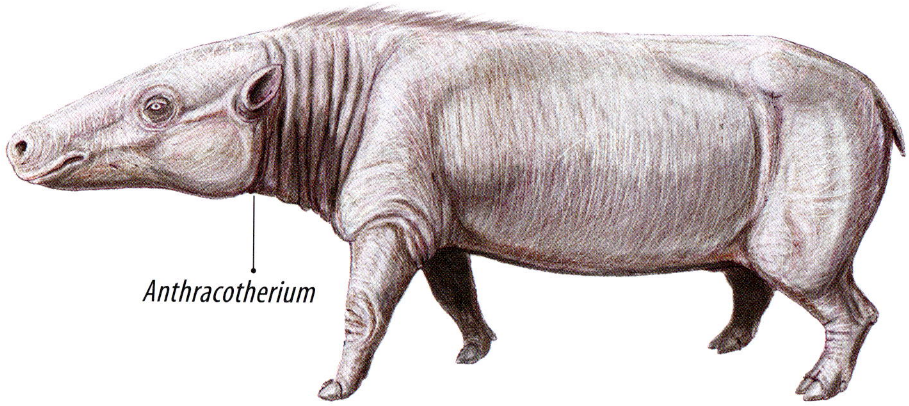

*Anthracotherium*

## What **lived then**

### LIFE ON LAND

*Hyracotherium* moved to the plains, and early tapirs and rhinoceroses continued to live in the tropical forests. During this epoch, the even-toed artiodactyls replaced the perissodactyls as the dominant medium-sized herbivores. The sheep-sized oreodonts thrived. Hoofed mammals flourished; the most famous included the *Anthracotherium*, known for having as many as 44 teeth. The largest land mammal ever, the hornless rhino *Indricotherium* was from this era. Also known as *Paraceratherium*, this giant animal was 18 ft tall, 8 ft long, and weighed about four times that of a modern elephant. South America had unique animals like armadillos, sloths and giant meat-eating ground birds. Other animals of the time included the tallest and heaviest entelodont the *Archaeotherium*, the Hyracodon which resembled a primitive horse but was actually a rhino, prehistoric horses such as the three-toed *Mesohippus*, and proto-camels such as the *Poebrotherium*.

South America hosted a bizarre array of mammals, including the elephant-like *Pyrotherium* that had three pairs of tusks, and the meat-eating marsupial *Borhyaena*, which had heavy, crushing teeth. In Africa, small-hoofed mammals called the hyrax evolved. A now extinct mammal class from the Oligocene was Embrithopoda. They looked like rhinos, but were close relatives of the elephant. The famous embrithopod was the *Arsinoitherium*, which had two large bony horns on its head.

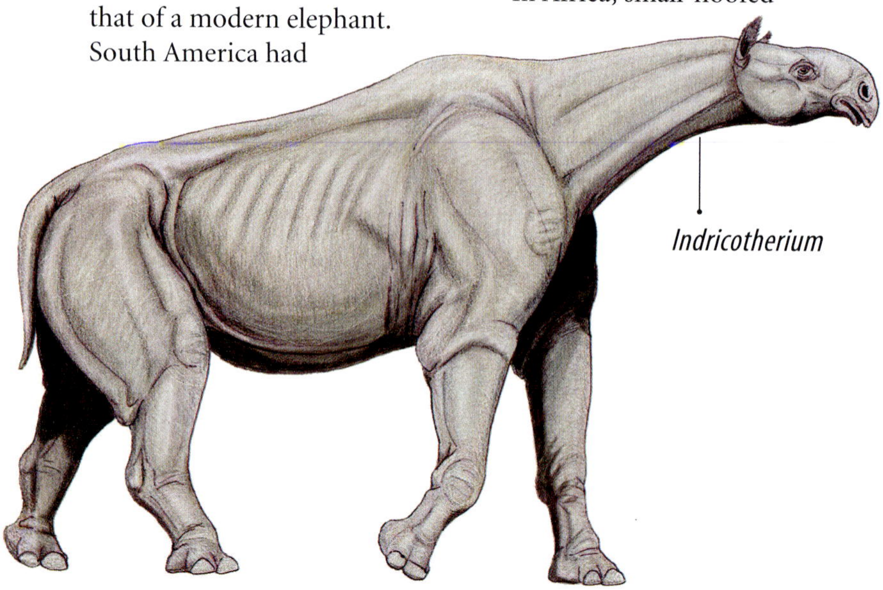

*Indricotherium*

### Did You Know

*Aegyptopithecus*, a small, tree-dwelling, fruit-eating animal that lived about 33 MYA, has been termed the "Dawn Ape". It weighed about 4 kg, and had certain similarities with modern-day lemurs, except that it had a full set of 32 teeth. The first ape-like form *Parapithecus* also hails from the Oligocene.

*Aegyptopithecus*

A recreation of Oligocene fauna in North America

## AVIAN LIFE

The giant predatory birds and penguins continued to rule. Towards the middle of the epoch emerged the *Psilopterus* – one of the smallest "terror birds". It varied in size from 70–80 cm to a little under 3 ft, and weighed between 5–7 kg. It fed on lizards, frogs, rodents, and large insects. About 27 MYA, the world's tallest and heaviest penguin appeared in New Zealand. Called the *Kairuku grebneffi*, it rose up to 5 ft and weighed about 60 kg. However, the most notable birds of the Eocene and Oligocene epochs were the giant, flightless, meat-eating birds, *Diatrymiformes*.

## MARINE LIFE

During the late Oligocene Epoch, seals, and sea lions appeared. Also, toothed and baleen varieties of present-day whales evolved.

## Causes of extinction

Global warming and massive carbon input to Earth's surface, resulting in rising sea levels and change in ocean circulation patterns caused the extinction of several species.

# NEOGENE **PERIOD**

The second of the three subdivisions of the Cenozoic Era is the Neogene Period. The word "Neogene" which means "newborn" emphasizes that the fossils of creatures from this period were more closely related to each other than to those of the "old" creatures of the preceding Paleogene Period. This period is subdivided into the Miocene Epoch (23–5.3 MYA) and the Pliocene Epoch (5.3–2.6 MYA).

## Time **zone**

- Began 23 MYA
- Ended 2.6 MYA

## What was the **Earth like**

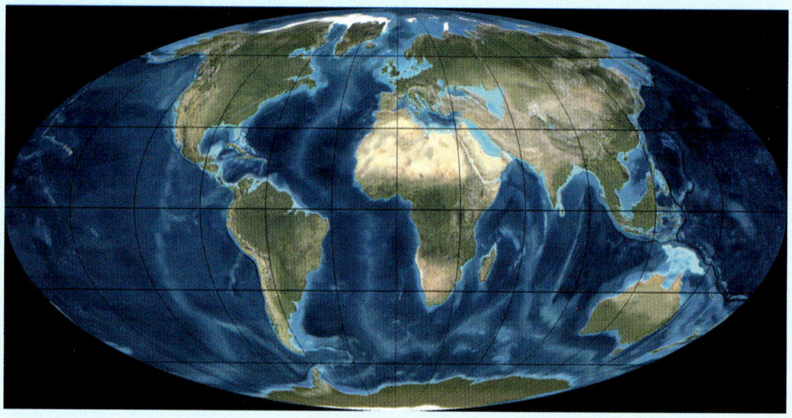

South America away from Antarctica led to the opening up of the Drake Passage between the two continents. As Antarctica became isolated from other continents, the circumpolar ocean circulation arose. It reduced the mixing of warm ocean water from the Tropics with cold water from the poles, resulting in the formation of the Antarctic ice cap. Simultaneously, the Arctic ice cap grew and became thicker.

### THE CONTINENTS

During this period, the continents moved closer towards their current positions. Australia continued to drift northwards, while the northward movement of Africa, Arabia, and India continued pushing the Alps and the Himalayas upward. South America moved away from Antarctica and merged with North America about 3.4 MYA with the formation of the Isthmus of Panama. This unification resulted in the separation of the Atlantic and Pacific Oceans. Moreover, the movement of

### THE MOUNTAINS

This period also witnessed the formation of the Alps, Pyrenees, Carpathians, Zagros, parts of the Rocky Mountains, Sierra Nevada, and the Cascade Mountains. With the emergence of these mountain ranges, remnants of the Tethys Sea were cut off from the Indian Ocean and got confined in the Mediterranean Basin. Today's Mediterranean Sea, which was formed at the beginning of the Pliocene, is the last remaining part of the Tethys Sea.

As snow fell on the high mountains and turned to ice, water was locked up far from the oceans, resulting in the lowering of sea levels. Consequently, land bridges came up between Africa and Eurasia and between Eurasia and North America. The Bering Land Bridge allowed faunal interchange between Asia and North America. However, towards the late Miocene, the bridge was flooded.

*Mont Blanc – the highest mountain in the Alps*

A recreation of the Neogene Period

## Did You Know

Fold mountains, the most common type of all mountains on Earth, form where two continental plates or a continental and an oceanic plate move towards each other. This movement forces sedimentary rocks on the sea floor to become crumpled and folded before they rise above the sea level as fold mountains.

Rocks folded upward are called anticlines, while those folded downward are called synclines. There are two types of fold mountains, viz. Young Fold Mountains (10–5 million years of age) like Rockies and Himalayas and Old Fold Mountains (over 200 million years old) like Urals and Appalachians of the United States of America.

The formation of fold mountains

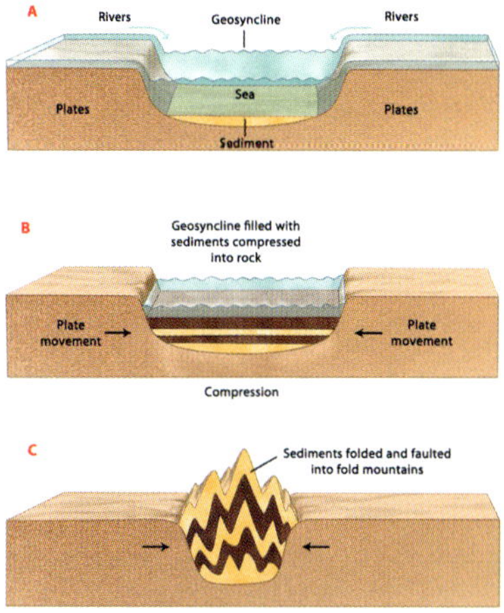

# MIOCENE EPOCH

The Miocene Epoch, the first subdivision of the Neogene Period, was one of the longest epochs of the Cenozoic Era. "Miocene" means "less recent" and derives its name from the fact that during this epoch, there were fewer modern animals as compared to the Pliocene.

## Time zone

- Began 23 MYA
- Ended 5.3 MYA

## What was the Earth like

Early on in the Miocene, the climate warmed up due to the formation of several mountain ranges, which changed air circulation and weather patterns and affected rainfall distribution. The dry conditions brought about frequent occurrence of drought in the rain-shadow areas, which favoured the expansion of grasslands. Well-defined climatic belts stretched from poles to Equator. However, the middle of this epoch witnessed global cooling, due to the expansion of the Antarctic ice caps and saw a drop in the concentration of atmospheric $CO_2$.

## What lived then

### THE GOLDEN AGE OF MAMMALS

Great diversification of mammals occurred and about half of the mammalian families known today were present then. Mammals took up new forms – from fast-running herbivores, large, predatory carnivores to small, quick-moving rodents. This diversity arose primarily because of

*It is believed that the first Native Americans travelled to North America via this bridge about 14 KYA.*

## Did You Know

During this epoch, the evolution of primates was also on the rise. From physical and genetic similarities, scientists have concluded that modern humans *(Homo sapiens)* chimpanzees and gorillas share a common ancestor that walked on Earth between 8–6 MYA. *Pliopithecus*, an early gibbon and a group of now-extinct human-like apes, called dryopithecines, were probably some of the earliest members of the lineage that eventually included humans.

The chief hominoids of this epoch were the *Kenyapithecus*, *Sivapithecus*, and the *Proconsul africanus*. The *Sivapithecus* was about the same size as a modern chimpanzee. Ancestor of the modern orangutan, it diverged from the common lineage of African apes and humans over 13 MYA.

*Sivapithecus*

*Chalicotherium*

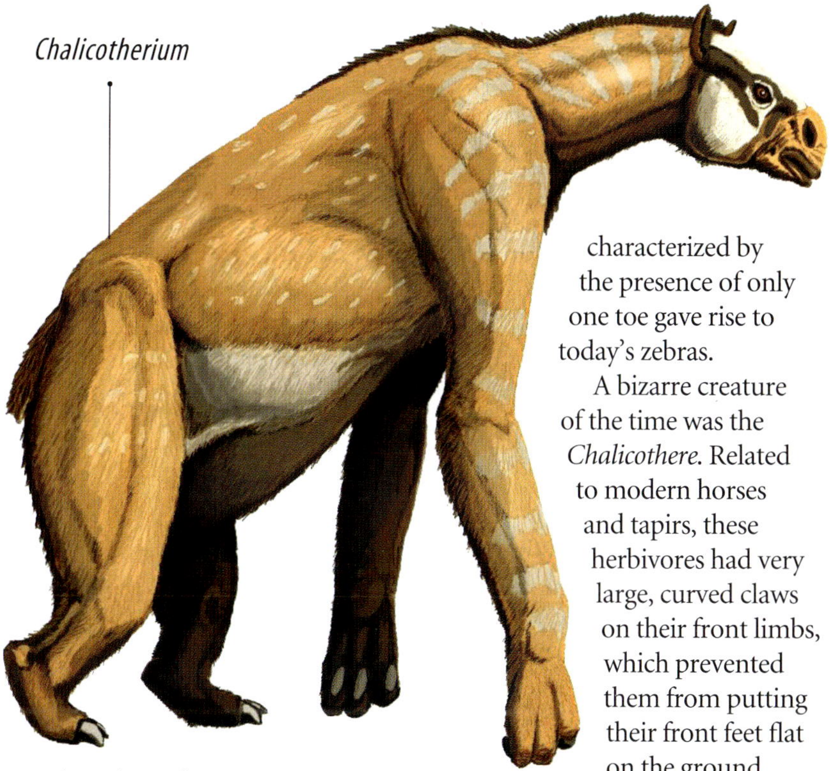

the formation of land bridges that allowed animals to migrate to new habitats. While elephants and apes moved out of Africa and settled in parts of Eurasia, rabbits, pigs, saber-toothed cats, and modern rhinos migrated in the opposite direction. Elephants and rhinos made their way to North America across the Bering Land Bridge, while horses moved to Eurasia. As North and South America came close to each other, ground sloths, which had evolved in isolation in South America, appeared in North America. Rodents reached Australia from Southeast Asia, after travelling along the Malaysian island chain.

## TERRESTRIAL MAMMALS

As grasslands spread in Eurasia and North America, giant camels that fed on leafy trees were soon on the verge of extinction. Though grasslands offered plenty of food for herbivores, they also led to certain changes in mammals. Towards the middle of this epoch, some animals developed durable teeth that helped them graze, while others developed high tooth crowns to tackle the tough dust on grasses. Horses, which had first appeared in the Eocene Epoch, began to diversify during this epoch, and grew in size. They had short legs, suited ideally for the forest-like land. Gradually, they developed longer legs and hooves that enabled them to run faster. Grazing horses with durable teeth diversified into many forms, out of which a lineage characterized by the presence of only one toe gave rise to today's zebras.

A bizarre creature of the time was the *Chalicothere*. Related to modern horses and tapirs, these herbivores had very large, curved claws on their front limbs, which prevented them from putting their front feet flat on the ground. So they walked on their knuckles, with their claws bent in towards the wrists.

## Fossil Fact

*Proconsul africanus*, a common ancestor of both apes and early humans, lived about 25–5 MYA in Eastern Africa in both forests and open lands. It weighed about 11 , had no tail, was strongly built, could climb trees, but was not able to swing from branches or leap from tree to tree. Unlike modern chimpanzees and gorillas, it did not walk on its knuckles.

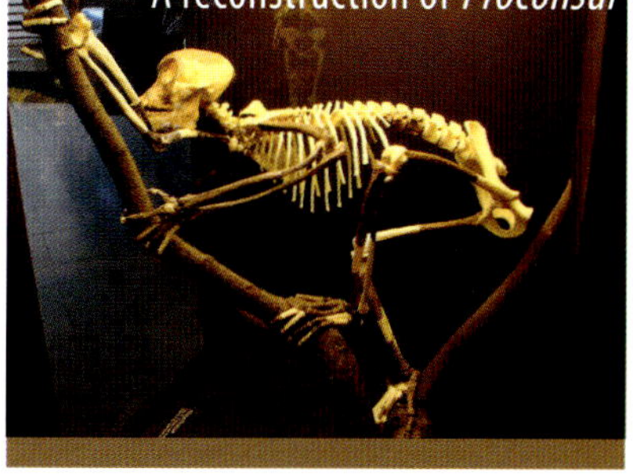

A reconstruction of *Proconsul*

Deinotherium

Ancestors of modern elephants, the mastodons, with their distinguishing tusks that curved upward and outward, spread to the Eurasian continent. From there, they migrated to North America before reaching South America.

Another distant relative of the elephant known as the *Deinotherium* appeared during this time. These gigantic animals had a short trunk and downward-curving tusks in their lower jaw as opposed to elephants that have tusks in their upper jaw. They flourished in the savannah regions of Africa.

Earliest ancestors of the antelope, the deer and giraffes appeared in Eurasia. The creodonts of the Eocene were replaced by more intelligent carnivores. The plains were populated by dog-bears – a bear that looks more like a dog – like the *Hemicyon*, which was found in North America and Eurasia. As opposed to modern bears, the *Hemicyon* was largely carnivorous, chasing down and killing primitive horses. This animal walked on its toes, unlike today's bears that have their entire feet on the ground as they walk.

The first hyenas appeared in this epoch, evolving from the earliest ancestors of civets. The period also witnessed the emergence of carnivorous marsupials in South America and Australia. An example was the *Nimbacinus dicksoni*, a small, fox-like animal that was a distant relative of the "Tasmanian tiger". About 20 inches in length, it fed on small mammals and birds.

## APES

As many as 100 species of apes lived during this epoch, including relatives of the orangutan, and the tree-dwelling *Dryopithecus*, which was found mainly in Europe. Like today's great apes, this creature had a short snout, strong jaws, and blade-like canines. Its thin enamel suggests it survived mainly on fruits. Its brain size was comparable to that of a chimpanzee. It spent most of its time swinging from branch to branch. On ground, it walked on all fours, like a chimpanzee. However, unlike chimpanzees, it used the flat of its hands while walking, rather than its knuckles. Another ape of the epoch was the *Oreopithecus* that roamed in Africa and Europe between

Orrorin tugenensis

### Did You Know

Scientists believe that fossilized remains of a piece of lower jaw, unearthed in 1997 in Ethiopia, followed by six teeth discovered in 2002 at Asa Koma belong to a new species called *Ardipithecus kadabba*. This bipedal creature lived in Eastern Africa between 5.8–5.2 MYA. Scientists who discovered the remains claim that it was a direct ancestor of man and is the oldest of all species belonging to the human family tree. The size of its brain and shape of its body were similar to that of a chimpanzee and it had larger and narrower teeth suggesting it ate fibrous food besides fruits and leaves. It probably used simple tools like twigs and sticks to gather food and stones to break open nuts.

Ardipithecus kadabba

11.6 MYA and 3.6 MYA and appears to have features of both the primitive monkeys and advanced human-like apes.

In 2002, the fossil of a great ape *Pierolapithecus catalaunicus* was discovered in Spain. It lived about 11.9 MYA, weighed about 35 kg, and fed on fruits. Like modern great apes and humans, it had a wide and flat rib cage. Its short and stiff lower spine suggests it might have been able to adopt an upright posture, and climb vertically in an ape-like manner. Like apes and humans, the *Pierolapithecus* had flexible wrists and short toes and fingers of monkeys. Since it had both ape-like features and some monkey-like characteristics, the *Pierolapithecus* is referred to as a "concestor", or common ancestor of humans and the great apes.

## THE EARLIEST HUMANS

*Orrorin tugenensis*, who lived about 6 MYA, lies at the base of the human family tree. However, it had more ape-like features than those of a human. Also known as "Millennium Man", its size was similar to that of a female chimpanzee, weighing between 30–50 kg. Its teeth were small, with thick enamel like that of modern man. Its diet comprised leaves, fruits, seeds, roots, nuts, and possibly insects. By studying its thigh bone, scientists have reached the conclusion that *O. tugenensis* walked upright on two legs, although it also climbed trees.

Another contender to claim the title of "earliest human" is the *Sahelanthropus tchadensis*, which lived about 7–6 MYA in Chad, Central Africa. Inhabiting both forests and grasslands, it walked upright and had a combination of ape-like and human features. For instance, it had a small brain, sloping face, elongated skull like that of an ape, and small canine teeth like those of humans. It also had a more evolved layrnx, which would eventually allow for more complex speech in humans.

## REPTILES

Although snakes, turtles, and lizards continued to diversify, the most notable reptiles of this epoch were gigantic crocodiles. The caiman-like *Purussaurus* of South America was one of the largest crocodiles, which grew between 36–43 ft in length and had large conical teeth to grip and hold onto its prey. The *Rhamphosuchus* was found in India. It probably resembled a modern-day gharial, but was much larger, 30–42 ft, or even bigger.

## AVIAN LIFE

Herons, ducks, eagles, hawks, crows, sparrows, pheasants, owls, and

Carcharodon megalodon

partridges were present in Europe. About 6 MYA, the world's largest flying bird, the *Argentavis magnificens*, took to the Miocene skies. It had a wingspan of over 20 ft and weighed about 68 kg. Since it was so heavy, simply flapping its wings would not have sufficed to let it fly. Hence, it glided over the plains of Argentina, using rising columns of air.

The *Pelagornis* had a serrated bony beak, a wingspan of over 16 ft, weighed about 30 kg, and was found all over the world. The *Osteodontornis*, with a wingspan of 18–20 feet, were bony-toothed seabirds that were probably ancestors of modern pelicans and storks.

## MARINE LIFE

As the epoch drew to a close, almost all modern groups of whales had appeared, as well as primitive seals and walruses. Marine crocodiles, sea lions, and sea cows were common. The *Desmostylus* was a strange-looking, heavily built, four-legged marine mammal that was found in the North Pacific Ocean. However, this creature, which used its cylindrical, tusk-like teeth to feed on soft, aquatic plants vanished by the end of the epoch.

It looked like a hippopotamus and is believed to be distantly related to elephants and sea cows. It spent a lot of time in water, but would return to the shore to take rest.

Desmostylus regularly entered the water, but it returned to the land to rest

The now extinct mega-toothed white shark, *Carcharodon megalodon*, preyed upon these marine mammals. The largest carnivorous fish of all time, it is believed to have reached up to 50 ft in length and weighed 50 tonnes! Its jaw was 7 ft tall and 6 ft wide and lined with 7-inch-long teeth!

Snails, crustaceans, and cephalopods were common, with many of these marine invertebrates, bearing close resemblance to their present-day counterparts.

## What grew then

Open vegetation systems, such as deserts, tundra, and grasslands expanded and forests diminished. By the end of this epoch, as much as 95 per cent of the modern seed plant families had evolved. The warming of the climate towards the middle of this epoch, followed by a period of cooling, led to the spread of coniferous forests. As the weather became more and more arid, tough scrub plants evolved as new form of vegetation. A major ecosystem that evolved during this epoch was kelp, a type of large brown algae. It grows in cool, shallow waters and attaches itself to rocks and corals to forms dense "forests".

# PLIOCENE EPOCH

The Pliocene Epoch was the second and final epoch of the Neogene Period. Therefore, the term "Pliocene" means "more recent". A relatively brief period, it lasted less than 3 million years.

## Time zone

- Began 5.3 MYA
- Ended 2.6 MYA

## What was the Earth like

### CONTINENTS

By this epoch, the continents had drifted toward their present positions. An eastward movement of the Caribbean tectonic plate led to forming of the Isthmus of Panama. It linked the continents of North America and South America about 3.4 MYA, while separating the Atlantic and Pacific Oceans. In North America, the Cascades, Rockies, and Appalachian Mountains, as well as the Colorado plateau continued to be elevated. The Sierra Nevada Mountains and Alaskan ranges were slowly beginning to form. In Europe too, many mountain ranges appeared, including the Alps. India's collision with Asia pushed up the Himalayan Mountains even further.

At the beginning of the Pliocene, the Mediterranean Sea was formed. About 5.6 MYA, its basin dried up and became completely isolated from the world's oceans. Then, over a period of hundreds and thousands of years, it totally evaporated. About 5.3 MYA, waters from the Atlantic Ocean cut through the Strait of Gibraltar and a massive flood swelled the Mediterranean Basin within a span of only two years. Known as the Zanclean Flood or Zanclean Deluge, it reconnected the Mediterranean with the oceans of the world.

### CLIMATE

Throughout this epoch, the climate became cooler and drier, though it was much warmer than today. This general cooling trend, which began during the

*Formation of the Himalayas has been cited as one of the possible reasons for the cooling trend, which arose during the Pliocene Epoch*

The spread of grasslands during this epoch led to the flourishing of grazing animals

Eocene, finally led to the Ice Age of the Pleistocene Epoch. Several theories have been put forward to explain the cooling. Variations in the amount of heat transported by the oceans and rising atmospheric concentrations of greenhouse gases have been proposed as some of the possible causes. Towards the start of this epoch, the North Polar ice cap formed and receded with climatic fluctuations, but towards the end, the ice caps in the Arctic became permanent. Also, it was during this time that Antarctica became a frozen continent. Towards the middle of this epoch, glaciers were present not only at the poles, but also at the middle latitudes.

## What **lived then**

### TERRESTRIAL MAMMALS

Gradually, the animals became larger and developed bigger teeth ideal for eating grass. Moreover, their legs became longer and stronger, thereby helping them walk long distances to new feeding areas and run fast escape predators. Hyenas, wolves, bears, and bison roamed the countryside. The 4.5 ft tall *Hipparion* looked similar to the modern horse. Huge camels like the single-humped, 11.5 ft tall *Titanotylopus* migrated from North America to Asia, while

## HOMININS: A PREHISTORY OF HOMO SAPIENS

The Pliocene witnessed the emergence of several bipedal hominins. In 2009, palaeontologists discovered the fossilized skeleton of man's earliest ancestor – a 3 ft 11 in tall female – weighing about 50 kg. The species was named *Ardipithecus ramidus*, nicknamed "Ardi". It lived about 4.4 MYA in the wet woodlands of eastern Africa and was a non-chimpanzee-like common ancestor of both man and ape. It was bipedal, though its big toe was designed to grasp branches. This is why scientists believe Ardi was a biped on the ground, but a quadruped (walked on its palms) while on trees. Its diamond-shaped human-like canines suggest that Ardi was an omnivore – surviving on meat, fruits and plants.

Two hundred thousand years after Ardi – between 4.2 and 3.9 MYA – another species called *Australopithecus anamensis* appeared, probably in forests and woodlands that grew around lakes of eastern Africa. While its ankle joint resembled that of a human (indicating it probably walked upright), its long forearms suggest it climbed trees as well. The heavily enamelled teeth suggest a diet of hard-to-chew food.

*Australopithecus afarensis* – between 3.85 and 2.95 MYA – was totally bipedal, as is evident from an 88-feet-long footprint trail in Laetoli, Tanzania that was discovered in 1978. The footprints, preserved in volcanic ash, reveal that these early humans walked the same way modern humans do. The most famous *A. afarensis* fossil is that of an adult female, named "Lucy". Found in 1974, the 3.2 million year old skeleton was almost 40 per cent intact. *A. afarensis* had both human and ape-like features. It had the flat nose, small brain (one-third the size of a human brain) and long arms with curved fingers of an ape, while the small canines and bipedalism were the human features. It was also probably similar to living apes in terms of diet, growth and development. Males weighed about 45 kg and were 4 ft 11 in tall, while females weighed 29 kg and were 3 ft 5 in tall. It lived in woodlands with patches of grass and ate a plant-based diet. When fruits were not available, it probably survived on insects and small vertebrates, such as lizards.

*Australopithecus africanus* – between 3.3 and 2.1 MYA – lived in Southern Africa. It resembled the *A. afarensis* anatomically and was bipedal as well, but *A. africanus* had a bigger brain and smaller teeth. Also, like *A. afarensis*, it displayed marked differences in size between males and females. The males were on an average 4.5 ft tall and weighed 41 kg, while females were 3 ft 9 in tall and weighed 30 kg.

giant prehistoric bears like *Agriotherium* – 9 ft tall – roamed in Eurasia, North America, and Africa. The *Smilodon* was one of the few sabre-toothed cats that survived till the end of the Pleistocene Epoch.

### Reptiles

Crocodiles and alligators disappeared in Europe as the climate became much too cool for these cold-blooded creatures. However, varieties of venomous snakes, including rattlesnakes that made their first appearance and preyed on rodents and birds, increased. In warmer climates like Australia, as plant-eating animals became progressively larger, so did their predators. The largest

Smilodon

unique flaring cheekbones, which were used either to store food or to attract mates.

The saltwater crocodile first appeared during this epoch, and it is the world's largest living reptile. The biggest known crocodile was 27 ft long and frequented the deep lakes close to today's Lake Turkana in Kenya. Called *Crocodylus thorbjarnarsoni*, it lived between 4–2 MYA, and quite possibly was capable of swallowing our early ancestors!

In South America, the aptly-named *Stupendemys* appeared, with a shell that was almost 11 ft in length and over 7 ft wide. It was the largest freshwater prehistoric turtle that ever lived. It hovered below the surface of water bodies and feasted on aquatic plants.

## MARINE LIFE

Pinnipeds (flat-footed mammals like seal, walrus, and sea otter) were found in various parts of the globe. The major marine predators were toothed whales, seals, and sharks. Towards the end of the epoch, the *Carcharodon megalodon* became extinct. Corals in the Western Atlantic and Caribbean underwent a period of transition, with some species disappearing from the face of the earth, while others appeared for the first time. This was the consequence of variations in temperature and change in food supply.

## AVIAN LIFE

The Pliocene Epoch marked the end of "terror birds" and other large, flightless, predatory birds of South America. One of the snake known from Australia, the *Liasis dubudingala*, hails from the Pliocene Epoch. It is estimated to have been about 30 ft long, and might have dwelled on trees. These giant pythons could easily kill and eat large prey, such as the *Euryzygoma*, which was a marsupial herbivore known for its

### Fossil Fact

Fossil name: Stupendemys geographicus
Site: American Museum of natural History
Stupendemys lived 5 million years ago in large bodies of water along the costal margin of South America.

A fossil of *Stupendemys geographicus*

### Did You Know

The North American three-toed mammal Hipparion - a horse - crossed the Bering Land Bridge and entered Asia and Europe. Hipparion carried its weight upon a single toe, which ended with a hoof. Other toes were also present upon either side of this toe, but they were so small in size that they did not even touch the ground.

Hipparion

*Stupendemys geographicus*

Titanis

last terror birds was the *Titanis* – the only terror bird to appear in North America, as a result of the Great American Faunal Interchange. This flightless, carnivorous bird stood over 8 ft tall, was heavily built, had highly developed eyesight, very long legs, and an enormous beak to rip its prey apart.

A pinniped

## What grew then

The cooling and drying had a significant impact on the vegetation. Deciduous trees thrived, while conifers and tundra covered much of the north. The dry climate resulted in the spread of massive grasslands and savannahs across all continents, except Antarctica. In many areas, grasslands replaced forests. Tropical forests, which had once been predominant across both hemispheres, were now limited to a tight band around the Equator.

## Animal migrations

The Pliocene was a time of great migration, owing to the appearance of new land bridges. The North American three-toed *Hipparion* horse crossed the Bering Land Bridge and entered Asia and Europe, while mastodons entered the Americas from Asia. The formation of the Isthmus of Panama ended the 60-million-year-long isolation of South America. Sixteen unique South American animals, including armadillos, glyptodonts, flightless predatory birds, opossums, porcupines, and giant ground sloths migrated into North America via the Panama land bridge. As many as 23 North American animals like cats, dogs, bears, camels, horses, tapirs, saber-toothed cats, mammoths, peccaries (pig-like mammals with dark, coarse hair), and deer took the opposite route. This event is known as the Great American Faunal Interchange. Animal migration between the continents led to a lot of competition, which, combined with the inability to adjust to a new environment, led to extinction of a high number of South American immigrant species. That's why today many North American species are found in South America, but only a few South American species are present in North America. Horses and mastodons migrated between Asia and North America across the Bering Land Bridge.

*The glyptodon was a creature with armoured shells that resembled armadillos, but weighed as much as a car*

# QUATERNARY PERIOD

The Quaternary Period is the final sub-division of the Cenozoic Era. Although it constitutes a relatively small part of the entire 4.6-billion-year history of the Earth, many significant events, such as several "Ice Ages", and most importantly, the evolution of our own species *Homo sapiens* happened in this period. This period, thus, is also referred to as the "Age of Humans".

The period is divided into two epochs – the Pleistocene Epoch, or the Great Ice Age, which started about 2,588 KYA and ended about 11.7 KYA and the Holocene Epoch, the current interval of geologic time, which started about 11.7 KYA.

## Time zone

- Began 2.5 MYA
- Continues until today

## What was the Earth like

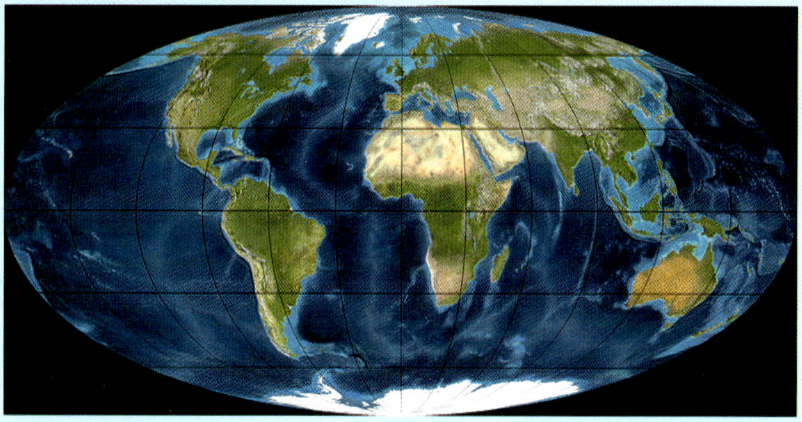

### THE CONTINENTS

The position of the continents hasn't changed much since the beginning of the Quaternary Period. In fact, the total continental drift has been less than 100 km; however, this shift has produced faults or large fractures – known as Quaternary Active Faults – in the Earth's crust. Geologists believe these faults can cause massive earthquakes in the future.

Throughout this period, a shift in the Earth's revolution around the Sun caused the arrival and departure of several ice ages.

When the ice retreated during the warm interglacial periods following an ice Age, mountains that had been re-shaped by glaciers were exposed. New rivers formed by the molten ice, flowing down these mountains, and drained into giant basins like the Great Lakes of North America. As glaciers advanced and retreated, coastlines and land surface areas changed considerably.

One of the most significant changes was the emergence of the Strait of Bosphorus, which now connects the Black Sea to the Sea of Marmara, and the Strait of Skagerrak, which lies between Denmark and Norway and connects the North Sea and Baltic Sea. With the formation of these straits, the Black Sea and Baltic Sea initially turned into freshwater bodies, but later, due to flooding caused by the rising sea levels, they returned to being saltwater seas.

As many as 20 cycles of advancement and retreat of

## Did You Know

The Milankovitch Cycle emanates from a theory about the long-term climate change, which occurs due to changes in the earth's orbit. Established by J A Adhemar and James Croll, the theory says a cyclical movement related to the Earth's orbit around the Sun causes deviations in Sun's heat received by the Earth surface. There are three cycles: eccentricity, axial tilt, and precession that impact climatic patterns over hundreds of thousands of years and are related to ice age cycles.

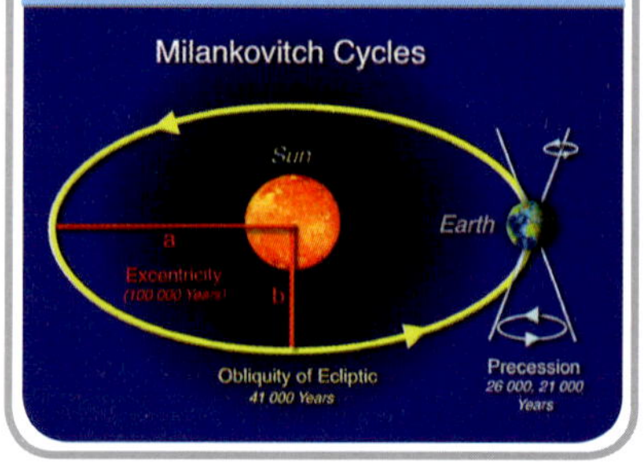

continental glaciers happened during the Pleistocene Epoch. Scientists estimate that during the most intense glacial periods, so much water was iced up in glaciers that sea levels diminished by over 400 ft, forming a tundra-covered land bridge – the Bering Strait – between Northeast Asia and North America about 70–11 KYA.

## Age of **humans**

### AUSTRALOPITHECINES

Scholars consider Australopithecines, a hominid family which includes a variety of species under the genus *Australopithecus* that evolved between 10–5 MYA as the earliest examples of human evolution. Studies related to the various species of *Australopithecus*, such as *A. africanus* and *A. afarensis*, have revealed that though possibly bipedal, these creatures walked in a way that was not identical to modern humans.

*A. afarensis*

### HOMO HABILIS

In 1959–60, the first fossils of a now extinct human species were discovered at Olduvai Gorge in northern Tanzania. Named *Homo habilis* in 1964, this most ancient representative of our genus *Homo* lived in parts of sub-Saharan Africa around 2–1.5 MYA. *H. habilis*, which means "handy or skilful man", got his name from the fact that simple stone tools were also

*H. habilis*

unearthed alongside the fossils, which led the scientists to believe that *H. habilis* was the first stone-tool maker.

Though *H. habilis* resembled modern humans more than the Australopithecines, it had a slightly bigger head, smaller face and teeth, human-like feet, and hand bones. It retained certain ape-like features, such as long arms and a protruding lower face.

### Tools

The Oldowan tools, meaning earliest stone tools, associated with *H. habilis* were first discovered by Mary and Louis Leakey at Olduvai Gorge.

An Oldowan tool

## Did You Know

A series of fossil finds provide evidence that modern humans first appeared in Africa some 200–150 KYA. Several species of ape-men co-existed in the Great Rift Valley of eastern Africa before migrating into Asia possibly between 2 and 1.8 MYA and entered Europe somewhere between 1.5 and 1 MYA. Migration into Australia, however, occurred possibly within the last 60,000 years and into the Americas only within the last 30,000 years or so.

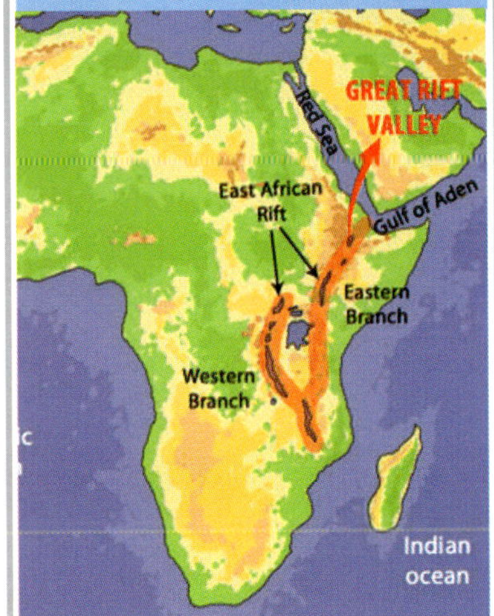

The Great Rift Valley

### Diet

The finding of animal fossils and bones alongside the Oldowan tools gave an idea that *H. habilis* were fearless hunters. They dragged and killed animals before butchering the carcasses and smashing open the bones. They used the stone implements to extract the marrow. The study of their teeth has revealed that they also fed on leaves, woody plants, brittle nuts or seeds, and very hard tubers.

## Language

Presence of a large cranium, or skull, indicates that they used some kind of a language, most probably grunts and primitive gestures, to communicate.

## Social Order

*H. Habilis* set up base camps where they lived in small groups. The males brought food through competition with other scavengers. The females gathered fruit, nuts, and roots. Thus, theirs was mainly a hunter-gatherer society with division of labour.

## HOMO ERGASTER

About 1.9 MYA, a fully developed human species, called *Homo ergaster*, which means "working man" appeared in Africa. *It* lived in eastern and southern Africa 1.9–1.4 MYA in the lower Pleistocene Period, which was characterized by global cooling. Initially *H. ergaster* used Oldowan stone tools, but by 1.6 MYA, it had learnt to make hand axes and cleavers that allowed better butchering of carcasses.

*An artist's rendition of an H. habilis at work*

## HOMO ERECTUS

About 1.89 MYA, another hominid species *Homo erectus* or the "upright man", originated in Africa – though some scientists believe *H. erectus* appeared first in Eurasia. However, during mid-Pleistocene (1.7 MYA), it had migrated to Europe, south Asia, and Southeast Asia.

In fact, the first *H. erectus* fossils were found in Java, Indonesia in 1891, giving him the name "Java Man". Similar fossils, which came to be called "Peking

*The H. habilis probably didn't have a language, so they only used simple sounds to communicate*

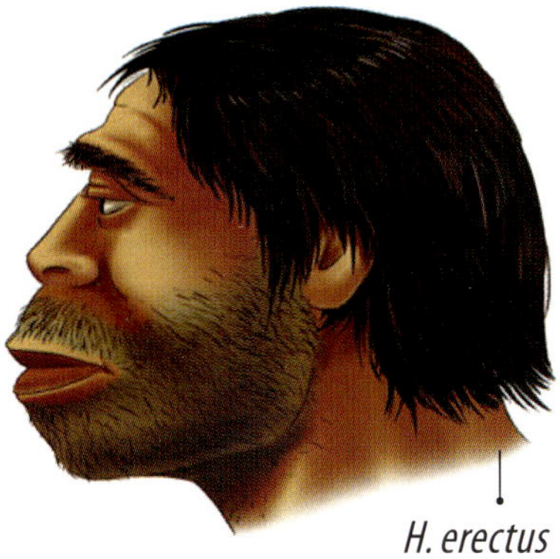

H. erectus

Man", were also discovered in 1927 near the village of Zhoukoudian in north China. The Peking Man fossils were not of a single individual but of men, women, and children. Subsequently, more *H. erectus* fossils were found at various sites of east, south, and northwest Africa, as well as Eastern Europe. From all these fossils, it has been concluded that *H. erectus* flourished till about 143 KYA, thus co-existing with *H. rudolfensis* and *H. habilis* when it first appeared, and later with *H. sapiens* towards the end of its time range.

## Appearance

As its name suggests, the "upright man" had a fully upright or erect stance. *H. erectus* was the first of our earliest ancestors, whose body proportions – with shorter arms and longer legs relative to its torso – were a lot like ours. These features indicate that *H. erectus* had developed the ability to walk, and even run, long distances. Since it has not been possible to differentiate between male and female fossils, it is generally accepted that its height ranged between 4 ft 9 in and 6 ft 1 in and weighed between 40 kg and 68 kg.

The brain of an adult *H. erectus* was about 69 per cent the size of a modern human's brain, and 50 per cent larger than that of *H. habilis*. Overall, *H. erectus* was a largely hairless, tall, thin, barrel-chested species with strong legs and thicker bones, features which made them a lot stronger than modern humans.

## Discovery of fire

Remains of ancient hearths and charred animal bones have led scientists to conclude that *H. erectus* was the first hominid to discover and use fire. It may have noticed it in nature - a volcanic eruption or a blaze that could have started by lightning. Later, it learned to conserve fire by blowing on the embers, and even later to create it at will by rubbing sticks or striking flints together.

### Fossil Fact

A 1.6-million-year-old skeleton of a 11–13 year old boy, known as Turkana Boy, was discovered near Lake Turkana in Kenya in 1984. The shape of his body indicates that this hominid was perfectly adapted to an active life under the Sun, leading to the conclusion that it was possibly the first to regulate its body temperature through sweating. It had a barrel-shaped chest, dark skin, and grew up to 6 ft 3 in in height.

A reconstruction of Turkana Boy

### Did You Know

Between 45–28 KYA, neanderthals began manufacturing more refined tools, known as the Châtelperronian culture. Fossils and artefacts unearthed in central and south-western France and northern Spain, between 1949 and 1963, revealed complex blade tools similar to those crafted by *H. sapiens*. Originally believed to have been made by modern humans, radiocarbon dating proved that these were, in fact, built by Neanderthals by imitating their contemporaries *H. sapiens*.

A model, depicting Châtelperronian culture

## Tools

*H. erectus* were accomplished tool makers and users. As far back as 1.8 MYA, they learnt to carve complex, specially designed stone tools with sharper and straighter edges. Their symmetrical tools were shaped like a leaf or teardrop, had pointed ends, and were sharply edged along the sides. About 1.5 MYA, their tools became sufficiently advanced and a new stone-tool industry evolved, called the Acheulian

*It is believed that the H.erectus controlled fire*

stone-tool industry, after the St Acheul site in southwest France. Its most important tool was the hand axe.

A stone axe

## Shelters

*H. erectus* were the first hominins to build a roof over heir head, under which they could sleep. Paleoanthropologists believe that *H. erectus* lived in small groups and moved frequently from one campsite to another, searching for food and avoiding bad weather.

## Diet

The ability to cook food paved the way for their long-term survival. *H. erectus* obtained much of the energy by eating meat. Roots, berries, shellfish, honey, and bird eggs were also part of their diet. A 450,000-year-old skull found in the cave of Arago in the Tautavel

Valley of France had traces of blows and marks on it, suggesting cannibalism, though the reason behind it remains a mystery.

## HOMO NEANDERTHALENSIS

*H. erectus* was followed by *H. neanderthalensis*, also known as neanderthals – after the Neander Valley, near Dusseldorf in Germany, where the first skeleton was discovered in 1856. They are closest to modern humans among all our extinct ancestors. Some scholars believe *H. neanderthalensis* was a distinct species of the genus *Homo*, while others say they were a sub-species of *H. sapiens* (*H. sapiens neanderthalensis*).

Neanderthals appeared between 300–100 KYA and lived till 35–24 KYA before the arrival of modern humans. They inhabited an extensive area – from today's England eastward towards Uzbekistan and almost to the Red Sea in the south – during the ice ages of the late Pleistocene Epoch, at the same time as modern humans (*H. sapiens*) were evolving in Africa.

### Intelligence

Although neanderthals are stereotypically depicted as dim-witted and brutish cavemen, research reveals an opposite picture. With brains as large as ours (and often larger), neanderthals had intelligent thought processes. They were skilled hunter-gatherers, who made and used a wide range of complex tools, built shelters, controlled fire, sewed and wore garments from animal hide, and even wore simple jewellery like pierced pendants! In fact, they developed and used a form of language, painted shells, and took care of the sick.

### Appearance

They were a physically diverse species, but they had bigger bones and more muscles than most modern humans. They were short and heavily built, reaching a height of 5 ft 5 in and weight of 65 kg, while females

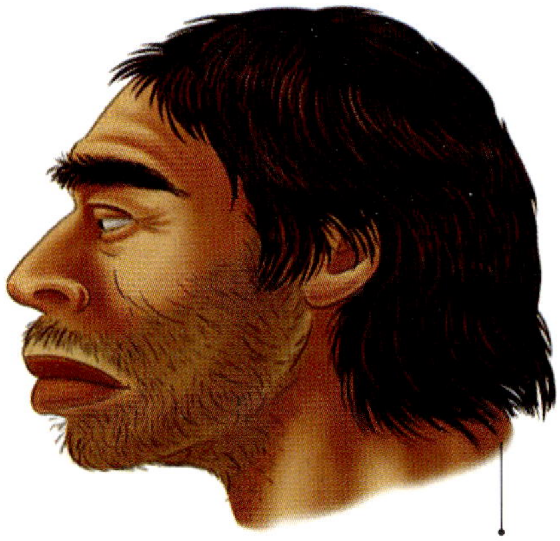

*H. neanderthalensis*

were, on an average, 5 ft 1 in tall and weighed 54kg.

They had broad shoulders, muscular arms, large chests, strong legs and hands, and broad feet like their ancestors. They had longer heads, their foreheads were sloped backwards as compared to the modern humans, and they had a prominent bulge at the back of their heads. Their large and especially long face had a big middle portion, a huge nose, and prominent brow ridges extending between the eyes, but no pointed chin. Thus the head and face of a neanderthal resembled that of a *H. erectus* more than that of the modern man.

### Tools

The neanderthals lived at a time when extremely cold climates had significantly reduced the supply of edible plant foods round the year. So they mostly relied on meat and became experts at being seasonal hunters. For instance, they killed reindeer in winter and red deer

Flint spearhead

**Did You Know**

Remains of a 750,000-year-old camp located on an ancient lakeshore have been unearthed at the Gesher Benot Ya'aqov site in northern Israel.

In 2000, on a hillside at Chichibu, north of Tokyo, Japanese archaeologists discovered the remains of the world's oldest artificial structure, a 500,000-year-old tent-shaped dwelling.

Huts and settlement similar to the one discovered in Japan were also found in 1966 at Terra Amata in France, dating from around 200,000 to 400,000 years ago. These probably housed a family or a group.

Gesher Benot Ya'aqov, Israel

in summer. Study of bone fossils and animal remains reveals neanderthals used sharp wooden spears to stab (rather than throw at) woolly mammoths, horses, deer, and other animals. When they lived in coastal areas, they also hunted seals, dolphins, molluscs, and fish.

## Diet

Neanderthals were primarily hunter–gatherers and ate a variety of food, including both cooked and raw meat. They hunted a wide variety of animals, including rhinoceros, bison, bear, and wild ass using their spears. They also ate nuts, tubers, as well as wild plants, which they knew how to cook. Those who lived in the Mediterranean regions fed on fish, shellfish, dolphins, and seals, besides hunting wild pigs, deer, and ibex.

## Language

Hunting tactics, tool-making skills, and other technical knowledge, such as extraction of bitumen, suggest that the neanderthals were capable of some form of communication. Structured speech was anatomically possible in neanderthals, because their brains had speech centres as large as those of modern man. Thus, it is quite possible that the neanderthals had a spoken language. The "language gene", called FOXP2, present only in modern humans, was also found in the DNA of neanderthal bones. This supports the claim that they were capable of speech and communicated with each other in their own language.

## Fire, clothing, and shelter

Reddened soil and remains of charcoal found at various sites show that neanderthals had learnt to control fire. They could produce sparks to start fires by striking flint with iron. It is also possible that they learnt to carry fire from one place to another. They built shallow hearths and used fire to keep themselves warm, cook, and protect themselves from wild beasts.

Neanderthals frequently occupied caves and rock shelters, whose openings faced the south, which ensured more light and warmth from the Sun, especially during the cold winters. Even though the term "cave men" is commonly linked with neanderthals, in reality, they also built temporary open-air shelters out of brushes or branches or made tents from animal skins and occasionally from bones of mammoths. Palaeontologists have discovered the remains of such a bone-shelter in Molodova, eastern Ukraine.

## Social life

Neanderthals are believed to have lived in small groups of 10–15 members, including children. The males were related to one another, while females moved in from other groups, possibly after marriage. They took care of the old and infirm. Study of bones above the tooth cavities of an elderly male, called La-Chapelle-aux-Saints-1, has revealed that other neanderthals chewed food for him, as he had lost all his molar teeth, before feeding it to him.

## End of neanderthals

The widely dispersed neanderthal population, at its maximum, probably comprised about 70,000 individuals. However, about 35 KYA, their numbers started falling. The last confirmed neanderthal site dates about 28 KYA, after which they seem to have vanished from the face of the Earth.

This abrupt disappearance roughly coincided with the appearance of *H. sapiens* in Europe, whose numerical superiority, more advanced brains, and better technological skills probably allowed them to prevail over the Neanderthals.

They might have been out-competed to extinction by the more adaptable *H. sapiens.* Research in 2010 suggests that the neanderthals became extinct

Neanderthals are our closest extinct human relatives

due to the most powerful volcanic activity in western Eurasia that occurred about 40 KYA.

## *HOMO SAPIENS* OR CRO-MAGNONS

Modern humans are classified as *Homo sapiens*, which means "wise man" in Latin. About 200 KYA, during a period of severe climatic changes, they appeared in Africa. Interestingly, they did not evolve from neanderthals, but were their contemporaries. But this is just half of the story. The remaining half leads us to the Cro-Magnons.

The first fossils of early modern humans were excavated in 1868 at a rock shelter in south-western France. The 30,000–32,000 year old fossils included the remains of four adult skeletons, one infant, and some bone fragments. They were called Cro-Magnon people, named after the cave they were found in.

The condition and placement of ornaments, including pieces of shell and animal teeth fashioned into what appear to be pendants or necklaces, led researchers to believe the skeletons had been intentionally buried within the shelter in a single grave. The site was one of the first to establish the ancient roots of modern humans, and fossils from this shelter represent some of the oldest *H. sapiens* populations of Europe.

### Physical features

Study of the fossils has revealed that the Cro-Magnons were a powerfully built muscular species, about 5'5"-5'7"tall. Unlike neanderthals, they had a smaller skeleton and lesser muscles, a tall, rounded skull, and the almost vertical forehead of modern man. They had a short and wide face, and they were the first humans to have a prominent chin. Their brain, however, was somewhat larger than that of the modern human beings.

Cro-Magnon

*It is believed that both* H. sapiens *and the neanderthals descended from* H. heidelbergensis

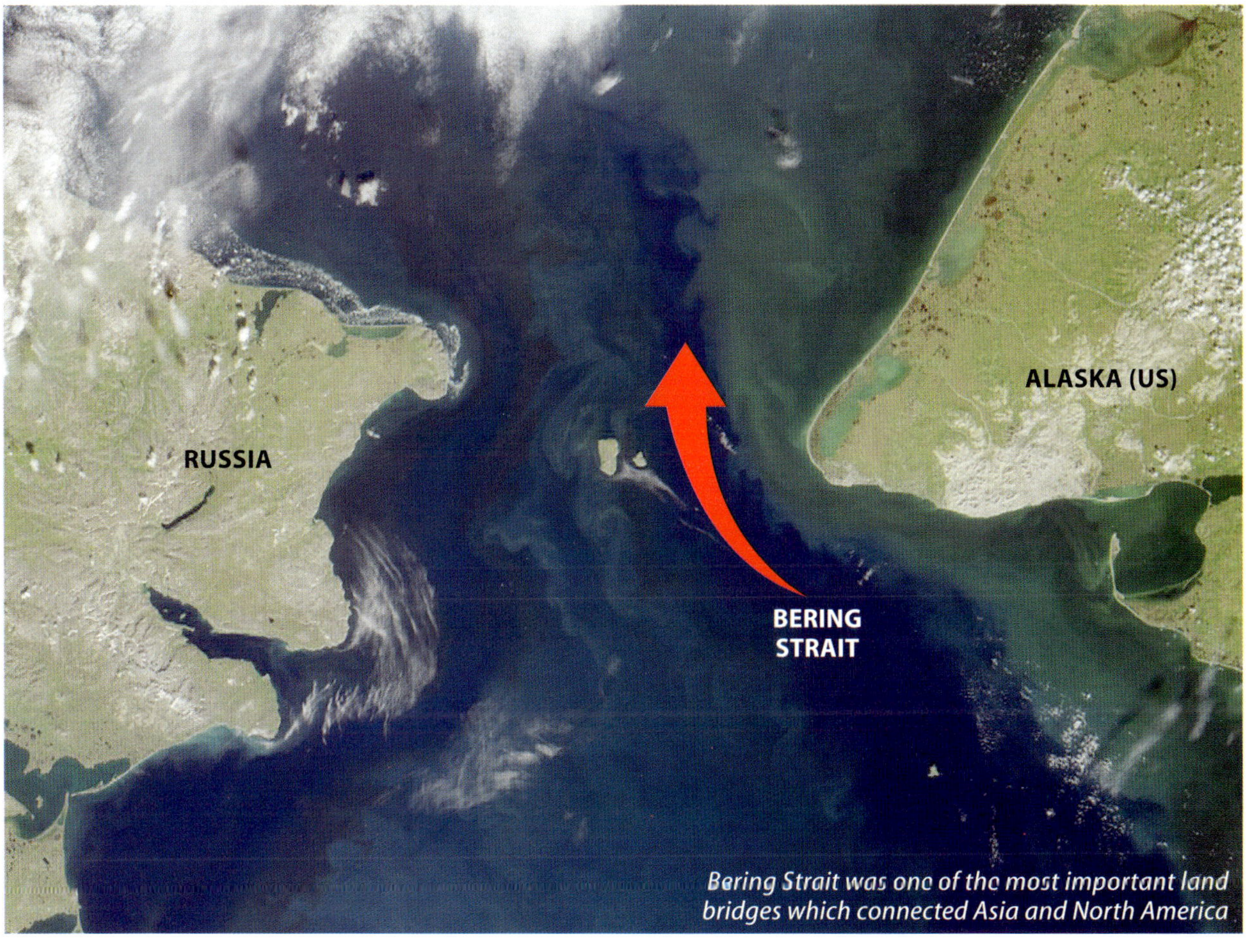


*Bering Strait was one of the most important land bridges which connected Asia and North America*

## Origin

According to the prevailing out-of-Africa theory – humans originated in and migrated out of Africa – modern humans evolved between 200–150 KYA, only in Africa. From there, they migrated to other regions about 40–60 KYA, and eventually replaced the existing populations of Neanderthals and other species that had descended from *H. erectus*. On the other hand, the multiregional evolution hypothesis also referred to as the regional continuity hypothesis – suggests that modern humans evolved in all major regions of the Earth almost simultaneously. It says that modern men of all parts of the world had a common ancestor – an early *H. erectus* who lived in Africa about 1.8 MYA – an extinct species of the *Homo* genus.

## Migration

It is believed that about 70 KYA, *H. sapiens* began migrating from Africa into East Asia, interbreeding with neanderthals in the process. Then between 60–46 KYA, *H. sapiens* travelled from Southeast Asia to Australia, possibly aboard simple boats or rafts. The Japanese Islands were inhabited about 30 KYA, and north-eastern Siberia between 35 30 KYA. Some *H. sapiens* entered North America via the Bering Land Bridge, situated between Siberia and Alaska, about 20–15 KYA.

## Shelter

During summer, Cro-Magnons lived in tents as they followed the herds they hunted. But in winter, they built permanent homes, which they used for many years. These homes were tepee-like huts, built from tree branches or bones of mammoths and covered with animal hide. Gradually tents gave way to mud-brick houses, coated with plaster.

At Çatalhöyük, in south-central Turkey, remains of a 9,000-year-old town have been excavated. As many as 10,000 people stayed there in rectangular mud–brick houses, which were built so close to each other that their entrances were possibly on the roof and were accessed by a ladder! Each house had a hearth, an oven, and platforms for sitting, sleeping, or working.

## Diet

Cro-Magnons were omnivores, and ate a variety of seeds, berries, roots, and nuts and hunted mammoth, reindeer, red deer, rabbits, hares, bison, foxes, wild horses, and tortoises. They also caught fish using bows and arrows, nets woven from vines, fish hooks, and even poisons. With the discovery of agriculture, about 10–7 KYA, they gradually figured out ways of curing, preserving, and storing food.

## Clothing and adornment

Cro-Magnons who lived in colder climates learned to make warm, comfortable clothes from animal fur by sewing together with needles made from bones and antlers and string from animal guts. Those in warmer climates made clothes from woven grass and even from tree bark.

**Did You Know**

The skeletons found at the rock shelter site at Cro-Magnon, near the village of Les Eyzies, France indicate that these humans people led a physically difficult life. Some of them had traumatic injuries and the remains of the adult female indicated she had a skull fracture. Their survival with such ailments suggests group support and care, which allowed their injuries to heal.

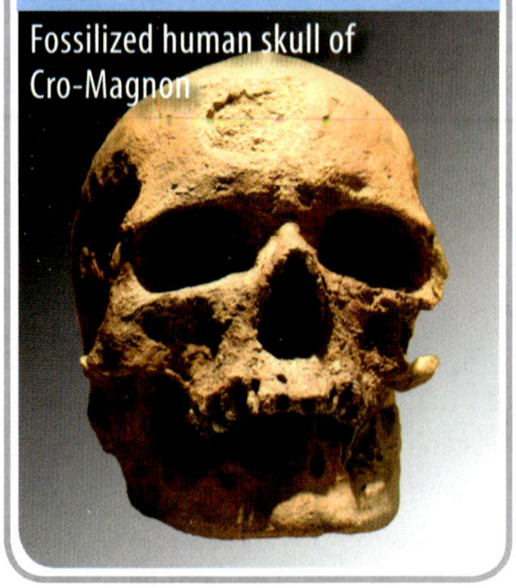
Fossilized human skull of Cro-Magnon

A model in a theme park depicting a Cro-Magnon hunting a mammoth

## Society

The Cro-Magnons lived in small tribes, consisting of multiple bands or family groups. Mostly, there were no class divisions in such societies. However, these were definitely hierarchical, with domestication of animals bringing about a dramatic increase in social inequality. Possession of livestock led to competition between clans and resulted in inequality of wealth.

Even though the Neanderthals were known to bury their dead, it was the Cro-Magnons who practised regular and elaborate burial. A 28,000-year-old burial site excavated at Sungir, Russia, contained the graves of three individuals dressed in clothes in which ivory beads had been sewn. Carved pendants, bracelets and shell necklaces also adorned the bodies. These "grave goods" suggest the Cro-Magnons believed in an afterlife.

## Agriculture

With the arrival of *H. sapiens*, the speed of progress and innovations increased dramatically.

About 9 KYA, for reasons that are still unclear, parts of today's Middle East underwent a transformation from a hunting-gathering society to an agricultural one. This revolutionary change originated at the Fertile Crescent, located between Armenia in the north and the Arabian Desert in the south. Because this transformation coincided with certain improvements in stone-tool technology, for instance, more precision in polishing, chipping, and grinding stones, this period (about 9–4 KYA) is also known as the Neolithic Period.

The early *H. sapiens* took to farming and manufactured a wide range of tools for tending, harvesting, and processing of crops. These included sickle blades, grinding stones, and bone implements for food production. But the most important tool that helped them clear forests on a large scale to create land for cultivation was the polished stone axe.

The advent of agriculture also redefined gender relations. Men

*H. sapiens made cave paintings, probably to record what they were leaving behind for the future generations*

took up the responsibility of cultivating crops, while women stayed back at home to perform household tasks.

Once dependency on agriculture increased, it became necessary for the neolithic people to live close to their farms, so they began to build permanent settlements with logs or stone. About 8 KYA, irrigation practices began almost at the same time in Egypt and Mesopotamia.

An Indian arrowhead

## Tools and Weapons

The Cro-Magnons crafted advanced weapons like stone axes, knives, spears, harpoons, wooden bows, and arrows with sharp stone tips, which allowed them to quickly adapt to and master their environment.

## Art

The Cro-Magnons were adept at sculpting small pieces of stone, bone, antler, clay, and ivory. The most significant of these sculptures are those of women, known as Venus figurines, which were crafted between 35 KYA and 10 KYA.

But Cro-Magnons are most famous for their astonishing paintings on rock walls, deep within caves. This form of cave paintings, which reached its pinnacle between 15 KYA and 10 KYA, has been found in over 150 caves spread across Western Europe.

## Technology

With an abundance of food and more permanent shelters, people could now devote more time to the development of new technologies. Better farming equipment, such as the ox-driven plough, was discovered. Invention of the wheel helped transport goods and people. The loom allowed weaving of finer and more comfortable clothing. Calendars, star charts, and sundials were created to measure the passage of time, which helped farmers track seasons and calculate the best time to plant crops. Gradually, people learnt to use fertilizers, to utilize irrigation better and deliver water from distant sources.

Discovery of metals, such as copper, brass, bronze, and iron, ushered in a new revolution in the history of mankind. Metals elevated technology from simple tools and weapons to today's computers and space travel.

# PLEISTOCENE EPOCH

The term Pleistocene comes from two Greek words, *pleistos*, meaning "most" and *kainos* meaning "new" or "recent". Pleistocene Epoch is, thus, the world's most recent period of repeated glaciations. It is also the earlier and more important of the two epochs that comprise the Quaternary Period of the Earth's history. It is characterized by repeated glacial periods that lasted about 100,000 years each time, when vast ice sheets covered much of the Northern Continents, and warmer interglacials of 10,000 to 15,000 years each, when the ice receded but did not completely melt.

The present Holocene Epoch, which began about 11 KYA with the retreat of the last continental glacier, is said to be an interglacial stage. It is very likely that the Earth will experience another glacial advance, perhaps in the next 10,000–20,000 years, and that the glacial or interglacial cycles will continue till they eventually come to a complete end and do not occur again for several hundred million years.

## Time zone

- Began 2.6 MYA
- Ended 11 KYA

## What was the Earth like

### CONTINENTS

At the beginning of this epoch, the continents were more or less in their modern position. However, slight movement of the tectonic plates formed the Isthmus of Panama. It affected oceanic circulation, and uplifted the Tibetan Plateau and broad regional areas of the western United States. This, in turn, affected atmospheric circulation, particularly the position and configuration of the polar jet stream.

In addition, the formation and melting of large ice sheets caused some vertical movements of the Earth's crust. This upward crustal movement continues today at a maximum rate of about 1.3 cm per year in the Hudson Bay area of USA, and in Fennoscandia, where the greatest depression and subsequent uplift related to the Scandinavian Ice Sheet is located in the Gulf of Bothnia.

### CLIMATE

During the Pleistocene Epoch, climate was not constant. Continental ice sheets advanced and retreated over large parts of North America and Europe. Earth experienced a number of prolonged glacial periods or ice ages and brief warmer interglacial phases like the one we live in today. The abrupt warming and cooling, when temperatures on ice sheets gradually cooled, then abruptly warmed by as much as 20°C, sometimes within less than a decade, shaped Earth's topography and soils and had a great impact on the life-forms.

### GLACIATIONS

In the Northern Hemisphere, there were four main glacial episodes, when ice sheets a mile or thicker advanced over northern Europe and North America. Radiocarbon dating of charcoal fragments has shown that the Wisconsin glaciation began advancing about 50 KYA, reached its maximum extent about 25 KYA,

## Did You Know

Glaciers move very slowly, flowing under the weight of ice, and they grow and shrink, depending on the climatic conditions. If there is a greater amount of snowfall and low temperature, then this will cause a glacier to grow as the snow is compressed into ice. In warmer periods, the ice can melt, forming icebergs and causing the glacier to shrink. It also moves by moulding like plasticine and by passing over a softer substance like sand or gravel.

Aletsch Glacier, Switzerland

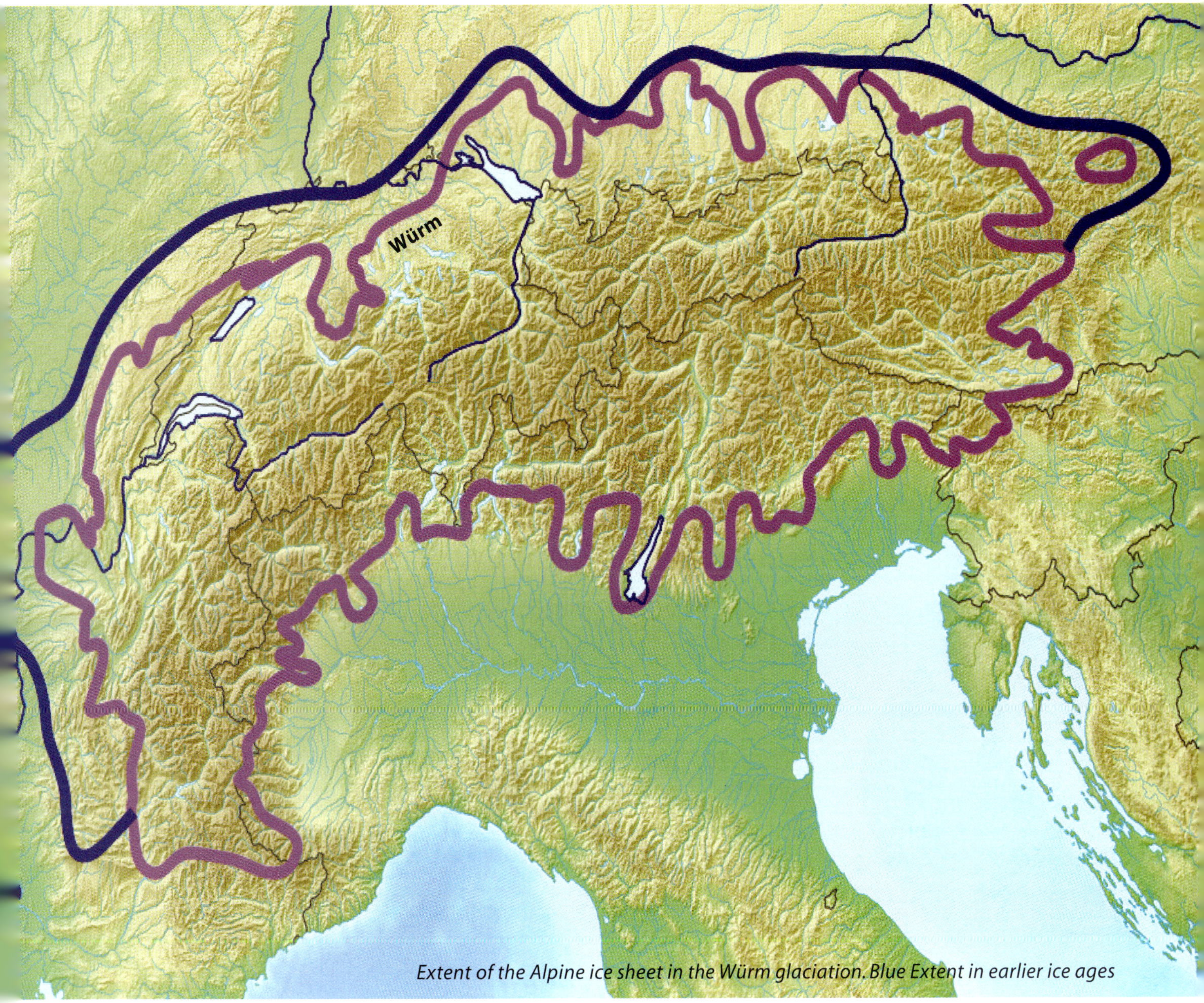


*Extent of the Alpine ice sheet in the Würm glaciation. Blue Extent in earlier ice ages*

and had retreated from Europe and most of North America by 10 KYA.

In fact, during the Wisconsin glaciation of North America, about 18 KYA, two massive yet independent ice sheets covered a large part of North America. Both the eastern Laurentide and the western Cordilleran ice sheets were over 3900 m thick. In Europe, during the Würm glaciations, ice covered Scandinavia, Germany, western Russia, and the British Isles. Another ice sheet covered most of Siberia. The glaciers of Greenland and Antarctica are remnants of the last glacial advance.

In the Southern Hemisphere, ice sheets had started to cover Antarctica from the preceding Pliocene Epoch and the continent was ice-bound throughout the Pleistocene Epoch. The Andes were covered by the Patagonian ice cap. There were glaciers in New Zealand and Tasmania. The present decaying glaciers of Mount Kenya, Mount Kilimanjaro, and the Ruwenzori Range in east and central Africa were larger. Glaciers existed in the mountains of Ethiopia and in the Atlas mountains. Since so much water was in the ice form, global sea level came down by approximately 140 m.

During the ice ages, global temperatures were 5°C cooler than today and it was much drier, as much of the world's water was locked up in huge ice sheets.

The Pleistocene is a unique epoch, because *H. sapiens* first appeared and migrated out of Africa during this time.

# What caused the Ice Ages

There have been approximately 11 ice ages on Earth. And even we are living in a mini ice age.

Ice ages seem to have occurred due to tens of millions of years of cooling. Once ice sheets began to grow, they probably contributed to their further development, as ice sheets reflected more sunlight back into space, which would have otherwise warmed the Earth.

The presence of large landmasses at high latitudes appears to be a prerequisite for the development of extensive ice sheets, because the large stock of ice or ice sheets cannot form over the ocean.

## CONTINENTAL BLOCKS

Movement of tectonic plates and uplift of large continental spaces causes profound changes in the global oceanic and atmospheric circulation patterns, which, in turn, result in climate change.

## GREENHOUSE GASES

Reduced quantities of methane and $CO_2$ may lead to global cooling. Many processes, including complex interactions among organisms, volcanism, ocean currents, and erosion can cause a long-term decrease in the amount of $CO_2$ in the atmosphere.

## EARTH'S ORBIT

The Earth's orbit varies through time. Important parameters that vary include its irregular

orbit around the Sun, the tilt of its axis, and the direction of the North Pole. Variation in these factors changes the amount and distribution of incoming solar radiation, which, in turn affects and initiates glaciations.

## EFFECT OF ICE SHEETS

Once the temperature cools, it's likely to stay cold. Ice has albedo, or reflectivity. A higher albedo results in less absorbed sunlight, because it's reflected back. This causes temperatures to drop more, so that the growth of one glacier triggers more glaciers.

## COSMIC DUST

Earth, along with the galaxy, rotates constantly in space. Probably the galaxy entered an area with cosmic dust, which blocked the Sun's radiation. This might have brought down temperatures on Earth.

## LOWER SUNSPOT ACTIVITY

A sunspot is a cool, dark spot on the Sun that contains magnetic energy. Sunspots keep the Earth warm, because their resulting magnetic field cuts cloud cover. With lower sunspot activity, cloud cover increases.

*The Pleistocene Epoch witnessed the emergence of megafauna animals or the ice age animals like woolly mammoths, giant giraffes, mastodons, and whales, who dispapeared at the end of the Quaternary Period*

## EFFECT OF GLACIATION

### Landscape

Glaciation during the ice ages played an important role in the shaping of landscapes in the middle and high latitudes. As the glaciers moved south, they scraped rock and soil in front of them. Soil and rock that was pushed in front of the glaciers commonly formed long ridges, called moraines. When glaciers retreated, blocks of ice detached from the main body. Depressions on the surface, called kettle holes, were formed when the ice melted. Large kettle holes that reached below the water table formed into lakes. Some kettle holes became wetlands, such as bogs, swamps, and marshes.

### Temperate regions

The formation of ice sheets over extensive areas killed several plant and animal forms, which could not adapt. Gaps and splits in rocks widened due to the wedging effect of ice, and riverbeds developed wide "U" shaped channels.

### Tropical Regions

Though ice formed on the upper reaches of mountains like the Nilgiris in South India, tropical areas largely experienced wet and dry climatic cycles. During the glacial periods, dry climatic conditions persisted as there was little cloud formation. Lowering of sea levels exposed large coastal shelves in tropical regions. The changing sea levels also affected the rivers as the erosional power of a river is increased due to lowering of sea level.

## What grew then

There were no major advances in plant types during the Pleistocene Epoch. In fact, growth of grasses and trees depended on falling and rising temperatures. Tropical jungles and rainforests were confined to the Equator, and the northern and southern regions were marked with deciduous forests, grasslands, and barren tundra.

## What lived then

### LIFE ON LAND

Although both marine and land animals during the Pleistocene Epoch were essentially like the modern animals, mammals of this period were much larger than their modern relatives. They had shaggy coats of hair as protection during the ice ages. They fed on small shrubs and grasses that grew at the ever-moving edges of the ice sheets.

Amongst the largest animals found during the Pleistocene Epoch was the woolly mammoth that lived in Europe, Asia, and North America. The mammoth belonged to the elephant family. It had 3 m long curved tusks and a prominent hump on the back. Mammoths that lived in cold climates had a shaggy covering of long, thick hair. These animals moved northward as the glaciers of the ice age receded. Like the mammoth, the woolly rhinoceros, an extinct species of rhinoceros, was common throughout Europe and northern Asia during the Pleistocene Epoch and survived the last glacial period. It became extinct too due to human and neanderthal hunting and climate change associated with the receding ice age. A bizarre South American hoofed mammal found during the end of this period was the 3 m long *Macrauchenia*. *Megatherium* was another popular giant sloth that once roamed this planet after the decline of the dinosaurs.

Woolly mammoth

*Macrauchenia*

The aurochs were the ancestors of domestic cattle, and they inhabited Europe, Asia, and North Africa. The caves of western Europe and Asia were dangerous places during the late

*Crocodylus anthropophagus*

Pleistocene Epoch as these dark, damp, and chilly dwellings were home to cave bears. They were also occasionally raided by hungry cave lions, who were found in Europe, Asia, Siberia, and America. The cave hyena ranged from Spain to northeastern China.

Animals in North America included short-faced bears, several species of tapirs and peccaries. Peccaries closely resembled the wild pig, had dark, coarse hair and a large head with a circular snout.

Mastodons are an extinct group of mammal species related to elephants. They inhabited North and Central America during the late Miocene or late Pliocene up to their extinction at the end of the Pleistocene Epoch 12 KYA. Animals of the cat family also included the American lion, *Miracinonyx* or the American cheetah and sabre-toothed cats like *Smilodon* and the scimitar-toothed cat, *Homotherium*. It was the most successful of all the sabre-toothed cats and was found in North and South America, Eurasia, and Africa. Other animals that also became extinct included the Irish elk, llamas, camelops, and giant polar bears.

South American wildlife in the Pleistocene also included a few grazers and mixed feeders such as the camel-like litoptern, and armoured mammals like *Doedicurus* and *Glyptodon*. Fauna in Australia was characterized by marsupials and monotremes, meaning mammals that laid eggs instead of giving birth to young.

## Reptiles

The Pleistocene Epoch saw mega-reptilian species like the giant monitor lizard *Megalania* and the giant turtle *Meiolania* of Australia and New Zealand, and *Crocodylus anthropophagus*, a large-sized apex predator reaching a length of 7–7.5 m. Other reptiles of this period included *Caiman venezuelensis*, a close relative of the current alligators and caimans that live in South America. *Colossochelys atlas*, an extinct species of *Cryptodira* turtles ranged from western India and Pakistan to Sulawesi and Timor in Indonesia during the dry glacial periods.

## MARINE LIFE

Since the outset of the Quaternary, the seas were ruled by whales and sharks. One fin-footed marine mammal that appeared on the scene during the Pleistocene was *Hydrodamalis*, a 10-tonne giant sea cow that became extinct 200 years ago. The giant shark, *Megalodon*, which had been the top predator of the oceans for millions of years became extinct during the Pleistocene Epoch.

*Doedicurus clavicaudatus*

## EVOLUTION OF HUMANS

The Quaternary Period is called the "Age of Humans." *H. erectus* appeared in Africa at the start of the period, and as time marched on, the hominid line evolved bigger brains and higher intelligence. The first modern humans evolved in Africa about 190 KYA and spread to Europe and Asia and then on to Australia and the Americas.

By the end of the Pliocene Epoch, a genus of extinct hominins *Paranthropus* evolved in Africa. Once known as robust australopithecines, the *Paranthropus* were bipedal hominids that probably descended from the australopithecine hominids about 2.7 MYA. The name *Paranthropus* comes from the Greek words *para*, meaning beside, and *anthropos*, meaning human. The name thus means "beside human" or "almost human" implying that paranthropines were not direct ancestors of modern humans.

*Paranthropus* was a very human-like ape. It had a slightly larger brain than *Australopithecus*. At about 5 ft height, it stood taller than its predecessors.

*P. aethiopicus*

*Paranthropus aethiopicus* was one of the earliest members of the genus *Paranthropus*. Its fossils found in Ethiopia in the Omo River basin in Kenya in West Turkana and Tanzania (at Laetoli) have been dated to between 2.7–2.3 MYA. *P. aethiopicus* fossil showed a relatively small brain size and a forwardly jutting face. It had large teeth, a powerful jaw, a well-developed sagittal crest – a ridge of bone running lengthwise along the midline of the top of the skull – and cheekbone, indicating it chewed hard foodstuff.

The first traces of *Paranthropus robustus* were discovered in 1938 in a cave in Gauteng, South Africa. The specimens recovered dated from 1.2–2.3 million years in age. Its brain size of was a bit larger than that of the typical chimpanzee.

*P. robustus*

*Paranthropus boisei* was an early hominin, described as the largest of the *Paranthropus* genus. It lived in eastern Africa during the Pleistocene Epoch from about 2.3 MYA until about 1.2 MYA. The largest skull specimen of *P. boisei* is dated to 1.4 million years, discovered at Konso in Ethiopia. Scientists named *P. boisei* "Nutcracker Man", because of its powerful jaw muscles, and its big and flat molars, which enabled him to eat nuts, seeds, and all kinds of hard fruit.

*P. boisei*

As you know, *H. sapiens* evolved from ancestors like those within the *Ardipithecus* Group, *Australopithecus* Group, and the *Paranthropus* Group. By 2.5 MYA, branching of a series of different kind of hominids had evolved in Africa. Clear lines separated Australopithicines and hominids representing the genus *Homo*.

The most primitive of the genus *Homo* is *H. habilis*. It is possible that *H. sapiens* are a direct descendant of *H. habilis*, who lived from 2.4–1.4 MYA. This species resembled modern humans, as it had a larger brain and smaller face and teeth. It got the nickname "Handy Man", because of its expertise at making and using tools.

Another member of the *Homo* group called *Homo rudolfensis* lived in Kenya 1.8–1.2 MYA. The first specimen of this species was discovered in 1972 in northern Kenya. In 1986, a Russian anthropologist gave it the species name *Pithecanthropus rudolfensis*.

*H. rudolfensis*

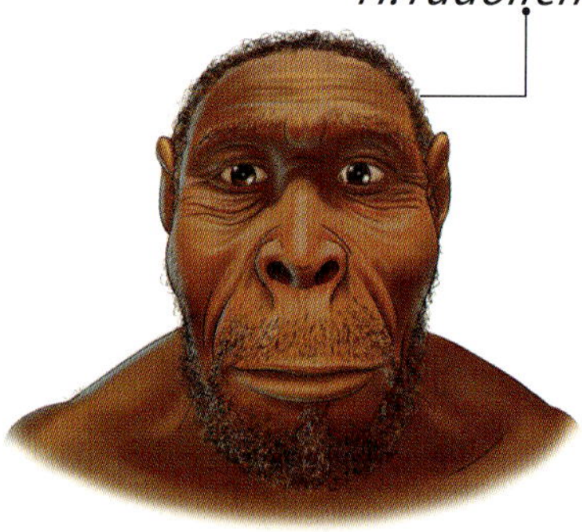

*H. rudolfensis* had very large teeth, but it lacked strong jaw muscles for chewing. Although no archaeological evidence was found with any *H. rudolfensis* remains, they could have also made some tools like their contemporaries *H. habilis*.

*Homo erectus* is the earliest known human ancestor to not have limbs that were used for climbing. It means *H. erectus* lived only on the ground and not on the trees, like its previous human ancestors.

The fossil of *H. heidelbergensis* was first found in 1908 in Germany and were dated to be 700,000–200,000 years old.

*H. heidelbergensis*

The most recently discovered fossils were that of an early human species, *Homo floresiensis*, nicknamed "Hobbit", on the Island of Flores, Indonesia. They existed 95–17 KYA. An almost complete skeleton and several bones of other individuals were found in a cave in 2003. They were small and had small brains. They made tools and hunted large prey. It is possible they knew about fire and used it to defend themselves against large predators. Though there were large animal bones found with their fossils, indicating the majority of their diet may have been meat, the large size of their teeth suggest they also often ate plants, fruits, or nuts.

*H. floresiensis*

## Did You Know

An ancient human relative *H. naledi*, found recently has been placed between the apelike Australopithecines and H. erectus. They are estimated to have lived between 2–3 MYA. Their hands seem suited to make basic tools and feet and ankles supported walking upright. Their curved fingers show they spent much of their time in the trees, and their teeth are surprisingly humanlike.

*A reconstruction of* H. naledi

## Evolution of the human skull

*About 4.4 MYA, a different kind of primates,* Australopithecines, *emerged. Belonging to the genus* Australopithecus, *these first hominids had not yet developed the large brain, teeth structure, and skeletal features, similar to the genus* Homo.

*Homo naledi, one of our ancient relatives, which lived between 2-3 MYA, had a tiny brain and therefore, a small skull. But its skull was round, with only a shallow slope down from the nose to the teeth. Its brow-ridge was also quite prominent, similar to that of* H. erectus

H. habilis *is the most primitive species of the* Homo *genus. This species resembled modern humans, as it had a larger brain and smaller face and teeth.*

H. erectus *appears to have evolved in Africa about 1.8 MYA and became extinct about 145 KYA. It was marked by an increased brain size, presence of brow ridges, a shortened face, and a projecting nasal aperture.*

*Neanderthals were our closest extinct relative. Their skulls included a large middle part of the face, angled cheek bones, relatively low, sloping foreheads, and a huge nose for humidifying and warming cold, dry air.*

*The skulls of* Cro-Magnons *reflected traits similar to humans, including the tall, rounded skull with a near vertical forehead. A large brow ridge no longer tops the eye sockets and there is no prominent prognathism of the face and jaw.*

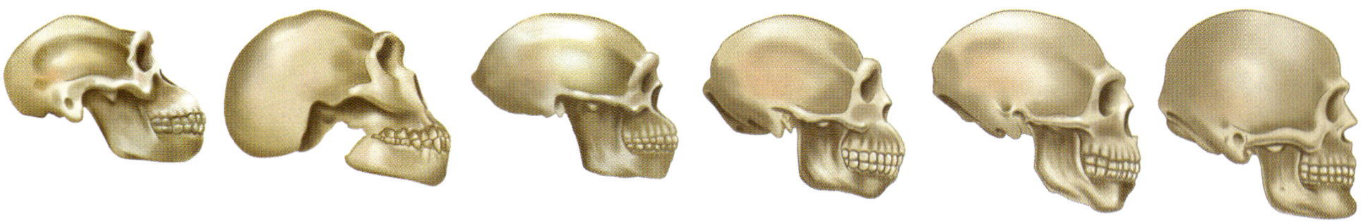

*Australopithecus* | *H. naledi* | *H. habilis* | *H. erectus* | *H. neanderthalensis* | Cro-Magnon

# HOLOCENE **EPOCH**

The present geological period, the Holocene Epoch, gets its name from the Greek words *holos*, which means "whole or entire" and *kainos*, which means "new". Thus the word Holocene means "wholly recent". The Holocene Epoch began 12–11.5 KYA at the close of the Pleistocene Epoch. The boundary between the earlier Pleistocene Epoch and the present Holocene Epoch is marked by global warming that led to the melting of the ice sheets that covered more than one-fourth of the Earth's land area. Since these ice sheets retreated at different times in different places, the date of the beginning of the Holocene is only an approximate boundary.

The present Holocene Epoch which began about 11 KYA with the retreat of the last continental glacier is said to be an interglacial stage. It is very likely that the Earth will experience another glacial advance, perhaps in the next 10,000–20,000 years, and that the glacial/interglacial cycles will continue till they eventually come to a complete end and do not occur again for several hundred million years.

### Time **zone**

- Began 11,000 YA
- Continues until now

## Post-Pleistocene **Earth**

### DEGLACIATION

The deglaciation of the Holocene Epoch turned the Earth's surface into its present form. The cooling of the Earth in the Pleistocene Epoch, which had caused the formation of ice sheets in the temperate regions had also affected distant arid and semi-arid regions. The increased rainfall in these areas resulted in the formation of large lakes in regions where rainfall was insufficient to establish a drainage system to the sea.

As ice began to melt in the beginning of the Holocene Epoch, sea levels rose up to 100 ft (30 m) as the glacial meltwater was returned to the seas. About 20 KYA, ice sheets as thick as 3 km covered

*The retreat of glaciers and change in sea levels marked the end of the Pleistocene epoch, and the onset of the Holocene Epoch, or the current mini ice age*

much of northern Europe, Asia, North America, Greenland, and Antarctica. The huge weight of this ice distorted and depressed the surface of the Earth's crust and the viscoelastic material from the Earth's mantle flowed away. As these glaciers melted, removing icea and creating space on land, it led to slow uplift or rebound of the land. This led to a return flow of mantle material and many areas rose as much as 180 m (600 ft) during the late Pleistocene and Holocene epochs. This process of uplifting of land is still going on, and it will take many thousands of years for the land to reach an equilibrium level.

*An artist's rendition of the Earth, depicting the ice age during the Pleistocene Epoch*

Scientists call this post-glacial rebound as glacial isostatic adjustment.

This post-glacial rebound caused a number of physical changes. In the Scandinavian region, it resulted in the formation of the Baltic Sea and weak earthquakes across Northern Europe. The equivalent event in North America was the rebound of Hudson Bay, as Tyrrell Sea shrank to its present boundaries.

Deglaciation of the Earth also resulted in a movement between the Earth's tectonic plates. It caused a tilting of the surface of the Earth as locations farther north rose faster than those on the southern edge. This can be seen clearly in the Great Lakes where the lakes are shallower in the north and deeper in the south. This tilting of the Earth also led to the formation of new rapids and rivers. Researchers have also found that as a result

## Did You Know

Extinctions in the present epoch are primarily caused because of deforestation, hunting, loss of habitat, etc. Once considered to be the most elegant and graceful of all species of kangaroo due to its beautiful fur and fanciful hop, traits which eventually spelled its demise, the Toolache wallaby likely went extinct by the 1940s. Hunted for its fur and for sport, the wallabies were often chased down using greyhounds.

Toolache wallaby

The Dryas Octopetala flower is known for growing in cold conditions

of the post-glacial rebound, the water mass around the Earth's surface is moving towards the North Pole at the rate of 0.035 in (0.88 mm) a year. This movement of the plate tectonics has caused an increase in earthquakes and volcanic activity.

## CLIMATE

The last Ice Age ended about 15 KYA as the Earth's climate began to shift from a cold glacial world to a warmer interglacial state. However, partway through this transition, temperatures in the Northern Hemisphere suddenly returned to near glacial conditions. This near glacial period, spanning 12.8–11.5 KYA , in the late Pleistocene Epoch, is called the Younger Dryas, named after the Dryas Octopetala flower, which was commonly seen in Europe during this period. Younger Dryas is characterised by cold and dry conditions increasing with latitude in the North, temperature and precipitation influences on tropical and boreal wetlands, Siberia-like winters in much of the North Atlantic, weakening of monsoon intensity, and southward displacement of tropical rainfall patterns. The Younger Dryas ended quite abruptly about 11.5 KYA when temperature began to rise from south to north. The rise in temperature was quite rapid and in Greenland, temperatures rose as much as 10° C in a decade. The Younger Dryas event is significant because it shows that even during an interglacial period, rapid climate shifts can still occur.

During the Pleistocene ice ages, as climate cooled and became wetter, glaciers grew in the Sierra Nevada Mountains. Rivers flowed into what are now dry deserts and lakes formed in many of down-dropped valleys of the Basin and Range.

The Holocene Climatic Maximum began roughly 9 KYA and ended about 5 KYA. During this period, the global temperatures reached their maximum level during the Holocene Epoch and the global temperatures were 1–2° C warmer than they are today. This period is called the Climatic Optimum or the Holocene Optimum.

As a result of the warming up, the north zones which, were uninhabitable until then, became habitable. On the other hand, large mid-latitude areas, such as the southwestern United States, which were previously productive, turned into deserts. It was during this period of Climatic Optimum that many of the Earth's great ancient civilizations rose and flourished.

In fact, about 6 KYA, the Nile River in Africa carried a huge volume of water. It was three times greater than what the Nile holds in the present age.

This Holocene Climactic Optimum was followed by a cooling trend that lasted from about 5 KYA to about 4 KYA. This cooling caused a large drop in sea

The Sierra Nevada mountain range, known for its glaciers, exists in the US state of California

In 2012, a cold spell across Europe led to the freezing of the Black Sea in many countries

levels. As a result, many coastal areas like the Bahamas emerged that are still above sea level today.

A short warming trend took place from 4–3.5 KYA, followed once again by colder conditions. This cold period, which lasted from 3.5–2.8 KYA, renewed the formation of continental glaciers and alpine glaciers, and once again sea levels fell between 2–3 m below present-day levels.

From about 2.8 KYA, the planet warmed up for a short period of about 600 years, although the temperatures did not get as warm as the Climatic Optimum. This was followed by another period of cooling that began from about 150 BCE and continued until about 900 CE. During this period, global average temperatures were significantly colder than today. In fact, at its height, the temperatures fell so low that the Nile River (829 CE) and the Black Sea (800–801 CE) froze!

During this period, the temperatures in Europe and the northern areas of North America were perhaps comparable to, or may even have been more than, those of the late 20th century. Grapes, fig trees, and olive trees grew in regions well north of their

## Did You Know

The Medieval Warm Period that lasted from the 10th -14th centuries, also called the Little Climatic Optimum, represents the warmest climate since the Climatic Optimum. So warm was the climate that the Vikings settled in Greenland and Iceland around 1000 CE and raised cattle and sheep until the onset of the Little Ice Age about 4 centuries later drove them back.

The Vikings' life

Olive trees have a long lifespan of about 2,000 years

Hvalsey church, in southwest Greenland, is the best-preserved artifact of Norse Greenlanders, who mysteriously disappeared in the 15th century

## Did You Know

The Little Climatic Optimum was followed by the Little Ice Age (also known as the Neoglacial Period) which lasted from 1550 to 1850 AD. During this period, global temperatures were at their coldest since the beginning of the Holocene – the average annual temperature of the Northern Hemisphere was almost 1° C lower than the present.

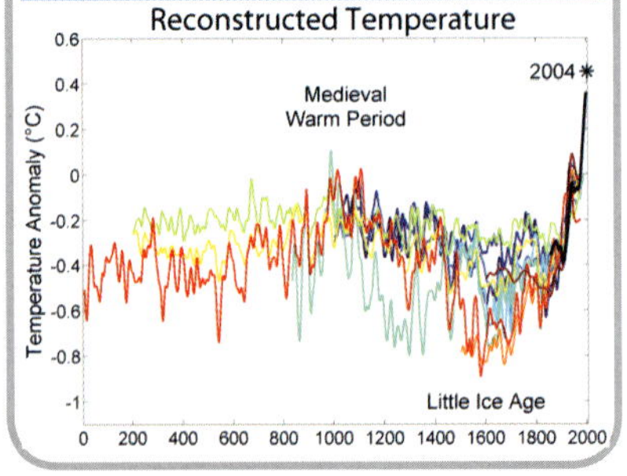


current range. In fact, even the snow line in the Rocky Mountains was about 370 m above current levels. These extreme weather conditions probably played an important role in the spread of the Black Death (bubonic plague) that killed an estimated 1.5 million people in Europe in the years 1348 and 1350.

However, the Little Ice Age was not continuously cold. While the 13th and 14th centuries were cold, conditions were more favourable during the 15th century, and then there was a return of more severe weather from mid-16th to mid-19th centuries. Western United States experienced one of its longest and most severe droughts between 1580 and 1600.

In Iceland, 25 per cent of the population was wiped out during 1753–1759, as severe cold weather destroyed crops and caused a famine. In New England, 1816 was termed "the year without a summer." Most scientists feel the recent global warming is the result of anthropogenic activities. They explain the cooling from 1940–1970 occurred as a result of sulphate aerosols formed by the sulphur dioxide gas, which released into the air by burning of "dirty" coal. Its aerosols cooled the Earth's surface by reflecting some of the incoming radiation from the Sun toward outer space. In the 1980s, regulations were passed to clean up the coal burning operations to reduce the amount of sulphur dioxide gas being released. As the amount of sulphate aerosols decreased in the atmosphere, the effect of greenhouse gases in increasing global warming could be felt once again.

On the other hand, the opponents of the theory stated above feel that these climate fluctuations are the result of natural causes. They cite two

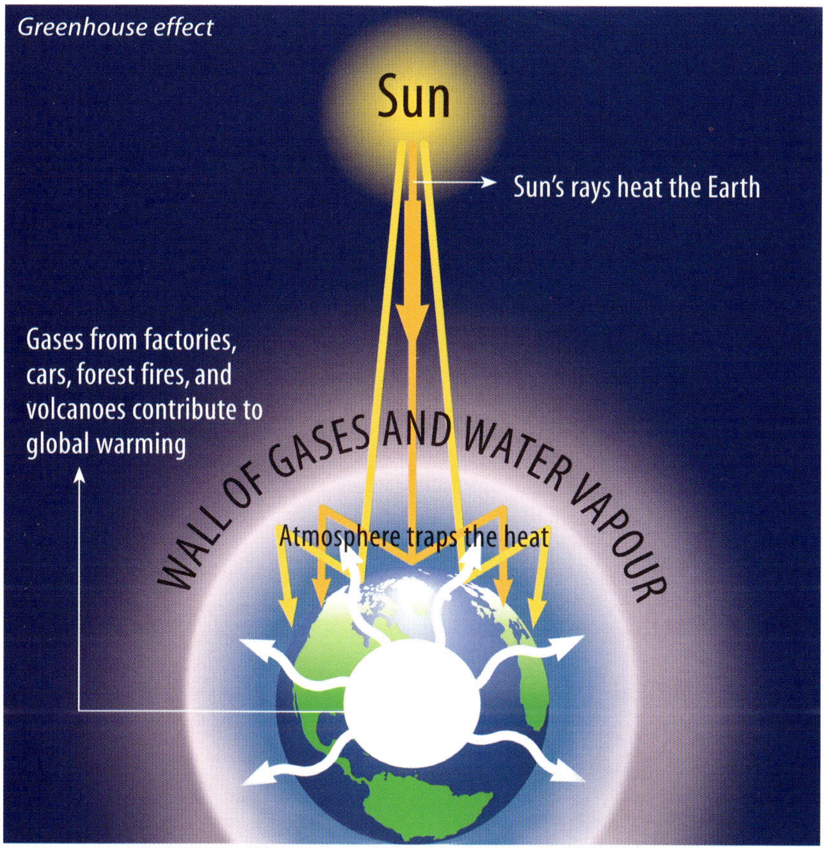


periods to prove their point.
a) The global warming during 1910–40 was too rapid to have been caused by the small increases in greenhouse gas concentrations that occurred during this period.
b) A global cooling occurred from about 1940 to the late 1970s even though greenhouse gas concentrations increased at the time. If the current global warming continues, a super-interglacial period might occur when the Earth could become warmer and the warm period could possibly last longer than any interglacial periods in the Pleistocene Epoch.

## Animal and plant evolution

While there has not been much evolution in animals and plants during the Holocene Epoch, there have been major shifts in their distribution. As Earth began to warm up, tundra was replaced by forests. Plant and animal species began to re-colonize the recently uncovered land. As the climate began to warm, some ice bridges that joined the continents melted away. As a result, some animal and plant species became separated from other populations.

While some species thrived in their new habitats, others died out. "Mega mammals" like the

*The Tundra regions are known for their extremely low temperatures, poor nutrients, short growing seasons, and treeless plains*

*Covering most of northern Africa, the Sahara desert is the largest hot desert in the world*

mammoth and woolly rhinoceros that had adapted to the extreme cold became extinct.

Human beings, who depended on them for much of their food, changed their diet to smaller hunting animals and plant materials. On the other hand, some animals became extinct in one region but survived in another, for example, horses and camels disappeared from North America but survived elsewhere.

## Humans and the **environment**

As the climate continued to get warmer, human population began to increase and we began inventing the processes that would change the planet forever. About 4.5 KYA, the previously lush and green Sahara, with its rich fauna, was reduced to dry sand.

The Moa, Tasmanian tiger, Dodo, Great Auk, and Passenger Pigeon of North America, all became extinct. Most of these extinctions are a result of human hunting practices and new diseases carried by humans and other new animals.

The rate of animal and plant extinction has been very rapid in the last century and a half due to the explosion in global human population and rapid technological development. Humans have moved into and over-exploited most areas of our planet, including the oceans. They have cut down forests, changed the courses of rivers, pushed wild animals and plants out of farm and urban areas, polluted wetlands with pesticides and other man-made chemicals, and hunted large land and marine animals. Tigers once ranged widely across Asia, from Turkey in the west to the eastern coast of Russia. Over the past 100 years, they have lost 93 per cent of their habitat. Similarly,

*A field of young corn plants indicating monoculture*

### Agriculture

Agriculture, involving domestication of plants and animals, developed about 10 KYA. It began in the Fertile Crescent, which includes Palestine, Lebanon, Syria, and Egypt. Cultivation in this area began when the warming of the climate drove out the reindeers, which constituted one of the principal game of the people of the area. Moreover, the vegetation of the cold climate was replaced by a Savannah, very rich in plants and cereals, leguminous plants, almond trees, and pistachio trees.

Gradually, the cultivation of wheat, barley, and other plants had spread from the Fertile Crescent through much of the Indo-European world. Domestication of sheep, goats, and cattle began at about the same time. In Central and South America, the most commonly domesticated plants were maize, bottle gourds, squash, and beans.

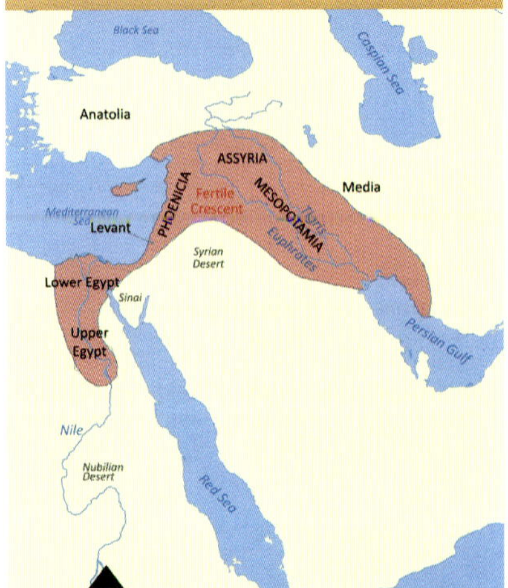


Fertile Crescent

An artist's impression of the Industrial Revolution in Ironbridge, England

a number of fish species have been depleted by fishermen during the last half century. Government efforts to save these wild animals from extinction have been only marginally successful. The World Conservation Union conservatively estimates that 7,266 animal species and 8,323 plant and lichen species are now at the risk of extinction primarily due to habitat degradation caused by humans.

Human activities have also had significant impacts on the geophysical characteristics of Earth. Monoculture – devoting large tracts of land to single crops – has affected the composition and fertility of the soil in most areas under cultivation. While the use of chemical fertilizers has improved soil quality, the harmful effects are still being felt.

**Did You Know**

A study by scientists at NASA's Goddard Institute for Space Studies shows that the average global temperature on Earth has increased by about 0.8 °C (1.4 °F) since 1880. In fact, since 1975, the rise in global temperature has been roughly 0.15–0.20 °C per decade.

## Humans and development

According to the theory of evolution, population expands until it reaches the carrying capacity, or the maximum number of individuals an environment can support without harmful effects, at which time it levels off – or stops expanding. Continued expansion beyond the carrying capacity generally results in a crash or a rapid decline to a level far below the carrying capacity. If enough genetic diversity remains, the population recovers, otherwise it becomes extinct.

By the first century AD, humans began to overcome these hurdles. There were approximately 170 million people on Earth at the end of the first century, by 1800 the population was over 1 billion. In the 19th century, the Industrial Revolution allowed human populations to grow exponentially. With industrialization, improved sanitation and medical caredeath rates declined.

At the same time birth rates too continued to climb in most parts of the world. According to the United Nations, world population reached 7 billion on 31 October 2011 and will reach 8 billion by 2024.

# THE LIVING PLANET

If you look carefully around yourself, you will be overawed by the incredible richness of life on Earth. This wide variety of ecosystems, ranging from coral reefs to polar tundra, and from tropical rainforests to temperate grasslands, from small birds to giant blue whales, is the result of almost three and a half billion years of evolution and extinction.

**Time zone**

- Began 11,000 YA
- Continues until today

## The rich fabric of life

A recent estimate suggests that there are about 8,700,000 species living today.

It is estimated that nearly half of the freshwater fish of South America have yet to be described and named. A recent exploration of deep-sea hydrothermal vents led to the discovery, not just of new species, but of entirely new families of creatures. In fact, even among our closest cousins, the primates, a member of the macaque family was found in the state of Arunachal Pradesh.Since 1980, about 40 "new" species of

*A tiny speck in the all-encompassing Universe, our breathtakingly beautiful planet contains life in all its spheres – air, water, and land*

A Royal Bengal tiger moving about in its natural habitat at the Sundarbans

monkeys have been discovered worldwide. The same is true for other species like birds, frogs, and plants. In an inaccessible part of the Foja Mountains, explorers discovered dozens, if not hundreds, of new species of plants and animals in what is probably the most untouched ecosystem in the entire Asia-Pacific region.

## BIODIVERSITY

This rich and varied web of life on Earth is called biodiversity, and the web of connections that binds it together is called the biosphere. The diversity of life forms on our planet are manifest not just in the ecosystems and species, but also in the vast number of genetically distinct varieties. For example, just as genetic differences exist between individuals, there are six subspecies of tiger because they evolved distinct characteristics as they became isolated from one another in different geographical areas. The large and powerful Royal Bengal Tiger is found in Bangladesh, Bhutan, India, and Nepal. The Indochinese Tiger that is found in several Asian countries such as Burma, Cambodia, Laos, China, Vietnam, and Thailand is smaller and darker than the Bengal Tiger. The tiger can only be found in the south of the Malay Peninsula. The smallest of all, the Sumatran Tiger, is confined to the Indonesian island of Sumatra. The Amur Tiger, found in Siberia, is large, pale, and thick-coated. The rarest, the South China Tiger, now no longer found in the wild, was once widespread in China. So you can see variety is found even within one species. Geographically isolated varieties of many other animals and plants show similar or even greater diversity.

### Did You Know

The loss of a butterfly or a bat can doom the plant it pollinates, and every tropical plant species that becomes extinct takes some 20 insects to extinction with it.

Butterfly sucking nectar

## Food Chain

At the bottom of the food web are millions of tiny single-celled organisms, a multitude of tiny worms, crustaceans and insects and a wide array of different fungi. The functioning of our planet and its ability to support this wide diversity of life is dependent on them. They recycle the Earth's organic waste, and in doing so, they release nutrients, enrich soils and enable the plant growth upon which all animals ultimately depend. They also help determine the composition of the Earth's atmosphere and are themselves at the base of many of the food webs that enable higher life forms to exist. But this biodiversity is disappearing before our very eyes. Indeed, the current rate of extinction has even been calculated to be proceeding at between ten and a hundred times faster than any of the prior mass extinction events in the history of the Earth.

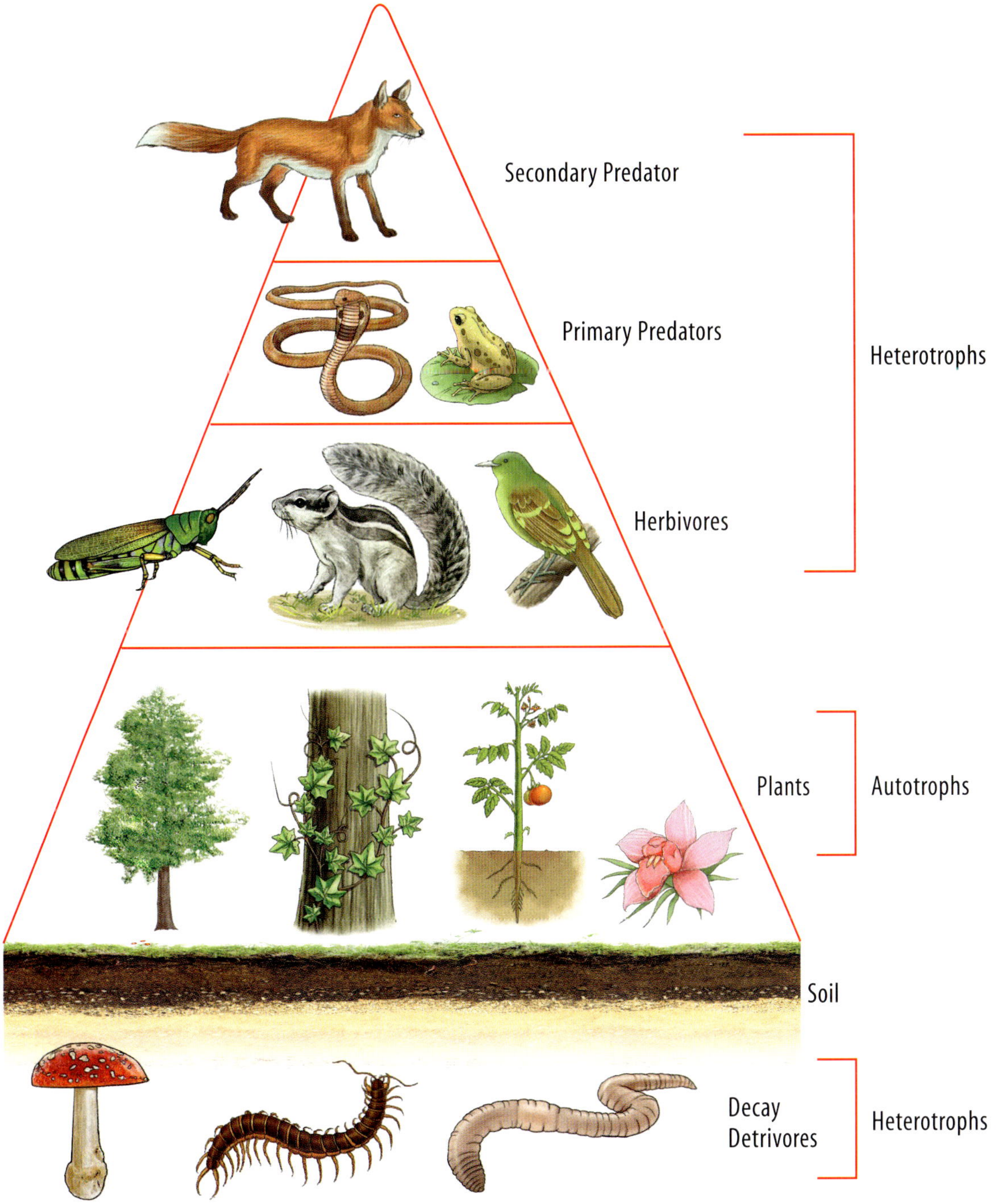

### Rainforests

While we often talk about loss of species, we have to realize it is most essential to preserve natural environments in order to conserve species and populations of animals and plants. Tropical rainforests, which lie between the Tropic of Capricorn and the Tropic of Cancer, are the most biologically diverse ecosystems on Earth. These tropical rainforests cover just six per cent of the world's land area, but account for perhaps half of the world's land species. And even within this relatively small proportion of the planet are areas where species richness is especially high. For example, although the rainforest in Colombia, Ecuador, and Peru is just two per cent of the Earth's surface, it houses over 40,000 species! Moreover, as compared to seasonal temperate forests or the colder boreal forests, tropical rainforests offer an incredible level of diversity of plants – of the quarter of a million or so vascular plants to have been classified, some 70 per cent grow in the tropical rainforests. For example, there are 780 tree species in a ten-hectare plot of Malaysian rainforest – more than the total number of tree species native to the whole of the USA and Canada.

## Threats **to life**

According to an estimate made in 2004, at least 784 species have become extinct since 1500 and another 15,589 are threatened with extinction – an estimate which includes one in four mammals and one in eight birds. The group that is most threatened of all is the amphibians – some 1,770 of the 5,743 known species are believed to be at risk of being lost. Experts put the real rate of loss at about one species per hour, and expect this to accelerate.

Less obvious than total figures, and less well-documented, is the disappearance of unique populations and subspecies. Each genetically distinct population of animals or plants has its own characteristics, some of these might be visible, such as differences in size or colour, others might not be so obvious, such as the ability to withstand cold or drought. Once the Earth's biodiversity is gone, it will be lost to us forever, never to be brought back. The rich fabric of our planet's biological diversity, woven over billions of years, will thus be torn apart, in what is for the life of the Earth, a mere blink of an eye.

### DEFORESTATION

The Earth is rapidly losing its forest cover. Compared to the year 1800, only about half of the tropical forests that then existed now remain. Between 1960 and

*A waterfall flowing through a thick green Malaysian rainforest*

*An aerial view of brown coal mining site, located near a vast forest expanse*

1990, about 4.4 million square kilometres (over 400 million hectares) of tropical forest were cleared – Asia lost almost a third of its tropical forest cover between 1960 and 1980; almost 90 per cent of West Africa's Upper Guinea rainforests have been destroyed. While there is no exact prediction of what will happen in the future, it is certain that if this continues, tropical forests will be reduced to between 10 and 25 per cent of their original extent by around 2100. The impact of that scale of loss of the Earth's biodiversity would be extremely serious. A satellite photograph taken in September 1987 showed a total of 7,603 fires burning in the Amazon river basin rainforests. Such images have sparked international concern and a number of initiatives have been taken to stem the destruction. While it is heartening to note that the rate of forest loss has declined, the continuing loss of these rainforests remains a cause for serious concern. The reasons for forest loss across this vast region are many and varied. Mining operations also have an impact both on tree cover and water quality. Clearing of large areas of forest for cattle-rearing, soya cultivation and palm oil plantations have also contributed significantly to deforestation. Logging operations and conversion to plantations, in turn, increases the risk of serious fires, especially when coupled with unusually dry conditions, which lead to further forest loss.

Tropical rainforests are not the only ecosystem under threat. Other forests, from subtropical to temperate to high-latitude boreal forests, are also under threat. Grasslands and heathlands are also being cleared for farming and urbanization. Even agricultural land that until recently maintained some level of biodiversity in field margins, watercourses, and small fragments of semi-natural habitats are suffering further depletion of biodiversity as farming methods become more and more intensive.

## DECLINING HABITATS

A number of species are being driven to the edge of extinction due to the rapid loss of habitat. In addition, fragmentation of habitat by farmland, roads, or cities is also affecting them. Small areas of wetland, forest, dunes, or grasslands are like isolated islands. If a particular species is lost due to disease or drought in one area, it is unlikely that it will be able to re-colonize from another area of suitable habitat. Animals that can't fly and plants that produce seeds that don't travel very far are at a greater risk.

Large carnivores which require large home ranges are especially

*Considering the speed at which we are cutting down trees, it is estimated that within 100 years there will be no rainforests*

vulnerable as suitable habitats are becoming more fragmented. Smaller areas of natural habitat are also strongly affected by climate changes. For example in the Amazon rainforest, fragmentation of the forest will disturb the present pattern of rainfall and the forest as a whole will thus become progressively drier. This, in turn, may increase risk of forest fires, leading to large scale degradation of the forest and its biodiversity.

Development pressures can also lead to important losses of habitat, especially where they are already fragmented or where small areas of habitat play an important role in conserving particular species. For example, the endangered Olive Ridley sea turtle have been placed at greater risk of extinction along the Orissa coast, as plans are currently underway to construct a massive deep water port at the mouth of the Dhamra River, just north of the Gahirmatha Marine Sanctuary, which is the largest of only three nesting sites for these turtles in the world. The turtles, already at risk from trawlers, gill netting, environmental factors such as global warming, and artificial lighting, will face additional stress from increased shipping traffic as it will deter the turtles from coming in from the sea to mate and nest. In addition, dredging, oil spills, and chemical leaks will add deadly pollutants to the ocean waters.

*Sometimes unintentional factors or natural factors like wild fires cause deforestation*

## MARINE ENVIRONMENT

Degradation of the marine environment has threatened

the world's coral reefs. These amazing marine ecosystems are often compared with the tropical rainforests for their incredible biodiversity. Approximately a quarter of all marine species inhabit coral reefs, and the number of individual species may be as high as a million. The greatest variety of coral is found on reefs in the area, in which the tropical Pacific and Indian oceans meet. There are some 400 known species of hard coral in the Philippines, 350 species of hard coral in the northern part of the Great Barrier Reef off the north-eastern coast of Australia, and about 250 further east into the Pacific Ocean near Fiji.

To thrive, corals need clean water at the correct temperature, and the algal plants that live with them need plenty of light in order to photosynthesize. Unusually warm water temperatures impair the ability of the algae to photosynthesize, revealing the coral's white exoskeleton. Corals can recover, if normal conditions return, but, if the stress continues, reefs can die off completely.

Activities carried out many miles from the coral reefs that discharge sewage water, agricultural pesticides, and fertilizers, soils that are swept into the sea as forests are cleared away, or as farmland is washed away by heavy tropical rainfall can drastically affect the health of the coral. Farm fertilizers and sewage encourage the rapid growth of algae suspended in the water, thus cutting down the light reaching the sea bed. Nutrients increase the growth of animals, such as sponges, thus not allowing the

*Uncontrolled fishing and other human activities pose a big threat to coral reefs in Indonesia*

**Did You Know**

Coral reefs are also plundered for corals for making jewellery or other decorative products, thus causing widespread damage to the reef.

Multistrand coral necklace

Crown-of-thorns sea stars are large, multiple-armed starfish, which feed on corals and can destroy entire reefs

corals to grow. Power-generation plants on land discharge extremely hot water that changes the water temperature, leading to coral bleaching and reef death.

Development and over-fishing has reduced the diversity of coral reef communities and has caused the coral reef ecosystems to become unbalanced. For example, more competitive organisms, such as algae, which were once controlled by large fish populations, have become more dominant. This is upsetting the whole system, leading to a much larger loss of species than simply the commercially valuable fish.

Moreover, as overfishing reduces catches, fishermen catch even the juvenile fish, thus causing even bigger population crashes as fewer and fewer of the fish make it to adulthood and breeding age. Use of explosives and poison by fishermen severely affect the long-term health of the reef as a whole besides killing all fish. Boat and ship anchors also cause widespread damage to reefs.

Sea defences, the development of large ports, and the expansion of large coastal cities have put other marine habitats also at risk. Structures such as groynes and jetties have eroded marshes. Oil and gas development on the island of Sakhalin off the far-

An image of Great Barrier Reef off the coast of Queensland, Australia

*A single tube of Aplysina archeri, a species of tube sponge, can grow up to 5 ft high and 3 in thick*

eastern coast of Russia threaten the extinction of the Western Pacific grey whale.

## UNWELCOME VISITORS

Biologists estimate that 10–20 per cent of established non-native species worldwide become invasive, causing serious economic or ecological harm. Aggressive animals and plants can wipe out more specialized creatures or plants with which they come into contact. In fact, introduced species have played a role in at least half of all the documented extinctions that have taken place since 1600.

One of the most amazing stories surrounding the introduction of a non-native species goes back more than a century, when Eugene Schifflin introduced starlings to America. Within 50 years, starlings had colonized the entire North American continent – with dire results. Starlings are noisy and live in flocks of up to hundreds of thousands, creating a noisy, smelly nuisance. They soil pavements, damage buildings,

### Did You Know

In 1890, a New York City man named Eugene Schieffelin imported to the USA every species of bird mentioned in the plays of William Shakespeare. Among the species that he introduced was a flock of European starlings.

Superb Starling

*Ancestor to all tigers, the South China tiger – the smallest of all the tiger subspecies – is the most critically endangered*

## Did You Know

A hundred years earlier, the extinction of this species seemed impossible, for the passenger pigeon was one of the most abundant birds ever to have existed. It is estimated that there were between three and five billion passenger pigeons at the time of European settlement of North America.

Martha – the last passenger pigeon

foul parks and playgrounds, and occasionally their droppings spread disease. In the countryside, massive flocks of starlings damage crops and consume feed meant for livestock. Further, these aggressive invaders drive out native bird species, particularly bluebirds, martins, and wrens from their favourite nesting spots.

As you are aware, Australia split away from the ancient supercontinent that was once the Earth's main landmass almost 200 MYA. This ecological detachment resulted in many unique life forms being found there and nowhere else today. For example, while the ancient marsupials disappeared across most of the rest of the world, they survived on the isolated island continent until the arrival of the Europeans. With human visitors came a whole host of other creatures, including camels, cane toads, goats, foxes, deer, rabbits, pigs, cats, dogs, and horses that are now at large in the wild in Australia. These newcomers compete with native animals for food, prey upon native animals, damage native plants, and degrade natural habitats.

Introduced species have had a dramatic impact on aquatic habitats too. The African Great Lakes – Victoria, Malawi, and Tanganyika – hold an incredible diversity of unique cichlid fishes. In Lake Victoria, however, the huge and voracious Nile Perch has become established and this single, exotic species may ultimately cause the extinction of most of the native fish, by feeding on them.

It's not just non-native animals that can cause problems,

*The Nile Perch is a massive freshwater fish found in Africa with a length of up to 6 ft and a weight of more than 225 kg*

Mountain bluebird

Sand martin

Fairywren

European starling

Water hyacinth plants have tremendous growth and reproductive rates

plants and introduced plant diseases can have a disastrous impact on native flora and thus the habitat of native animals. For example, Lantana, which is originally from the American tropics, was introduced into India at the East India Botanical Gardens, Calcutta, in 1807 as an ornamental plant by the British. But this plant escaped from the gardens and started spreading around the cities and from there it entered the forests. In fact, it is estimated that Lantana has invaded more than 130,000 km of area in India. It can prevent the young plants of the native vegetation from growing by releasing chemicals into the environment that are toxic to other plants. Finally, once it has established itself, it grows as clumps and prevents anything else from growing. It has become one of the biggest threats to conservation as instead of there being multiple kinds and types of trees and plants that a variety of animals can use, now the forest has mainly Lantana.

The Saint Johns River is Florida's longest river. In the 19th century, the Saint Johns was a busy waterway for the greater part of its course. The water hyacinth, a native of South America, a beautiful aquatic plant with purple flowers, was accidentally introduced into the region in the 1880s, and it spread with alarming rapidity throughout the upper reaches of the river. Now dense mats of water hyacinth render navigation almost impossible on the river's upper course. The plant is very difficult to eradicate, and it has also clogged the channels of other Florida rivers.

A reconstruction of Steller's sea cow – an extinct herbivorous marine mammal

## HUNTED TO DESTRUCTION

On 1 September 1914, at the age of 29 years, Martha died. Her death marked the final demise of the passenger pigeons. "In the autumn of 1813, I left my house at Henderson, on the banks of the Ohio River, on my way to Louisville ... The air was literally filled with pigeons; the light of noonday was obscured as by an eclipse; the dung fell in spots, not unlike melting flakes of snow, and the continued buzz of wings had a tendency to lull my senses to repose", writes John James Audobon in his Ornithological Biography. Once pigeon meat was commercialized and sold as cheap food, mostly for slaves and the poor, began an unregulated hunting campaign on a massive scale, which finally led to their extinction.

Since those far-off days, there has been a steady stream of hunting-derived extinctions. Some of the animals hunted to extinction include the Dodo, the great Auk, and Stellar's sea cow. The tiger's situation has grown desperate in a mere century. A hundred years ago, there were over 100,000 in the wild, with more than 40,000 in India alone. Today, the total number of tigers worldwide is calculated at less than 3,500. Three subspecies, Javan, Bali, and Caspian tigers, vanished during the 20th century. A fourth, the South China Tiger, has not been seen in the wild for more than 25 years and is assumed to have become extinct during the 1990s. Remaining populations include 1,850 Bengal tigers and a few hundred each of the Siberian, Indochinese, Malayan, and Sumatran subspecies. The reason for the dwindling numbers in India and elsewhere is the black market in the animal's body parts. Although the World Federation of Chinese Medicine Societies have removed tiger bone from their medicines, skins still sell for up to $35,000, and organs and body parts – bones, whiskers, eyeballs, paws, claws – are sold as souvenirs or as ingredients of traditional Asian medicine. Tiger is occasionally served at restaurants in Hanoi, Beijing, and Russia.

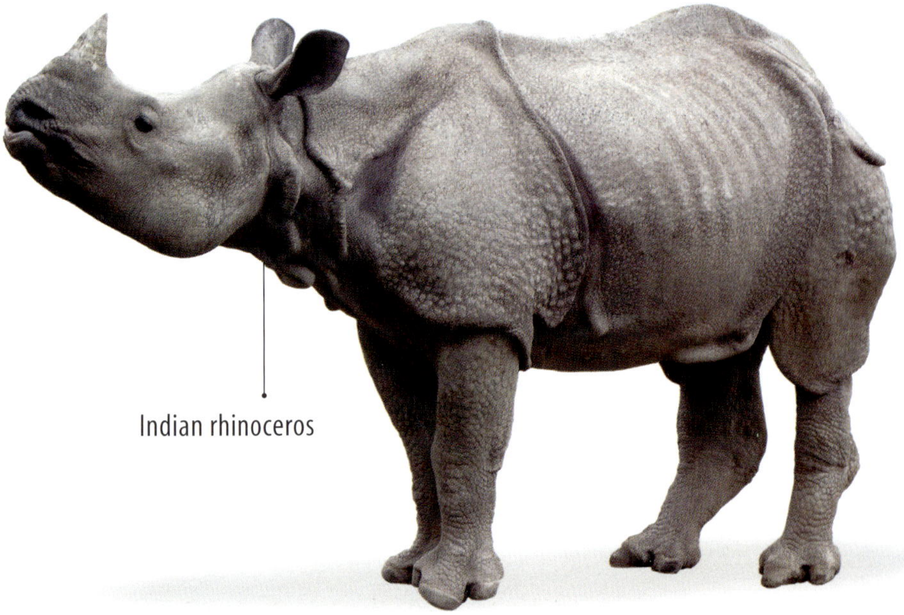

Indian rhinoceros

## Did You Know

Of the endangered species, 28 per cent of listed bird species and 29 per cent of mammals are threatened mainly by hunting. Although there are laws in place both nationally and internationally to protect rare and declining species from the most serious impacts of commercial exploitation, they are regularly violated as the hunger for food, furs, hides, horns and shells continues to place a great many species under serious pressure. In fact, the illegal trade in protected wildlife is so huge that it has been estimated as the third-largest illicit market in the world after drugs and weapons.

Confiscated skins and bones

This is a tiny population and these are slow breeding animals, but even they were increasing painfully from an even lower point, until their rise was set back by hunting. Between 1986 and 1995, about 450 rhinos were poached in India and about 50 in Nepal. In Manas National Park in India, it has been estimated that poaching has recently killed more than 95 per cent of the rhino population.

## PESTICIDES

Pesticides have been used to control various kinds of agricultural pests, and thus boost yields and make food cheaper for consumers. Some pesticides are designed to kill insects, for example, by attacking their central nervous systems, others are manufactured to target plants or kill microorganisms. They are not only used in the farmed landscape, but urban areas and gardens too.

The use of powerful pesticides can contaminate soil, water, turf, and other vegetation. In addition, they can be toxic to other organisms, including birds, fish, beneficial insects, and non-target plants. Some chemicals are highly stable and can persist in the environment for long periods. DDT is one such substance. For decades, it was successfully applied in large quantities to protect crops, but it did not immediately break down and travelled through food webs. DDT thus built up in the bodies of animals, accumulating in particularly high concentrations in top predators, especially in birds of prey, such as the bald eagle. This caused eggshell thinning, leading to decreased reproductive rates. In humans, DDT is linked with cancer, endocrine disruption, and reproductive and developmental ill-effects.

The most widely used pesticides today are organophosphates and carbamates that do not bioaccumulate and degrade rapidly in sunlight. But they are much more acutely toxic even at very low levels. They impact wildlife mainly by affecting the nervous system, leading to nerve exhaustion and, ultimately, failure of the nervous system.

## FERTILIZERS

Artificial farm fertilizers have also had a major impact on biodiversity. When artificial

Eutrophication poses huge problems to marine habitats, such as lakes, as it can cause algal blooms

nitrogen fertilizer is used in agriculture to stimulate crop growth and increase yields, a lot of the fertilizer doesn't stay in the fields. There is now an unnaturally high level of biologically-active nitrogen cycling in the atmosphere, on land, and in lakes, in rivers, and in the sea. This is having profound effects, causing farming-related nutrient pollution problems around the world. These environmental changes include a process called eutrophication. This occurs when nitrogen, and other nutrients, fertilize a water body and stimulate a "bloom" of microscopic algae. The rapid algal growth in turn reduces oxygen levels in the water to the point where animals, such as fish, die. Nitrogen pollution can also cause acidification of freshwater and terrestrial ecosystems and can change the nature of habitats, causing some species to be lost as more aggressive plants survive and thrive on the extra fertilizers

## DRUGS

An unbelievable 99.9 per cent of India's vultures have died out over the last 20 years, after becoming victims of unintentional poisoning. It took years to find out why India's three vulture species – the long-billed, slender-billed, and oriental white-backed vulture – were dying off. It turned out they were feeding on dead cattle and other livestock that had been treated with an anti-inflammatory drug called Diclofenac. It was discovered in 2004 that diclofenac causes renal failure in vultures.

## INDUSTRIES

Pollution arising from industrial sources has considerably impacted wildlife. Sulphur and nitrogen compounds released from coal-fired power stations in particular have led to "acid rain". This has destroyed forests across large parts of central Europe and has affected insects, fish, and sensitive plants, leading to the ecological demise of thousands of lakes in Scandinavia. India too faces an increasing threat from acid rain. Automobile exhaust fumes are partly to blame, but the worst culprits are coal-burning thermal power plants and steel industry.

Vulture eating carrion

While some ecological changes have been closely documented, others remain as yet poorly researched and understood. What we do know, however, is that the combined impacts of pollution, habitat destruction, introduced species, and the direct exploitation of wildlife is leading to a mass extinction of species on a scale not witnessed on Earth for millions of years.

Greenhouse gas emissions from a coal power plant

# SAVING PLANET EARTH

It has taken life 3.5 billion years to emerge in its present unbelievable richness. However, today, the finely woven fabric of life on Earth is seriously at risk. Habitat loss and the other pressures have brought much of the planet's biodiversity to the brink of disappearance or extinction. Ecosystems are getting degraded and fragmented due to pollution. The ecological relationships are getting affected by the impact of aggressive aliens and disintegration of habitats.

**Time zone**

- 11,000 years ago
- Present day

## Need for action

Should we be bothered about what is happening to the natural world? After all, as many argue, the manner in which we have used nature is justified by the huge improvements in the material well-being of many people. However, the scale of change now taking place because of our activities could mean that we pay a huge price ourselves.

### FOOD AND RAW MATERIALS

Although it is sometimes not immediately obvious, nature provides immense riches and benefits for humankind. For a start, everything that we eat is derived from once-wild species – cereals, vegetables, fruits, meats and everything else that we require for our dietary welfare is provided by species that have wild ancestors. At present, compared with the vast range of biodiversity that exists, we consume only a tiny handful of species. It may be quite likely that we have not identified all of the potential sources of food that might exist in the ecosystems that we are presently wrecking with such abandon.

As we cope with challenges like the rapidly growing population, climate change and the continuing challenge of ending poverty, it would be sensible for us to preserve as much choice as we possibly can to draw on in the future.

This is not simply a case of being able to find new food species that might help us adapt to future conditions; it is also important to protect the wild relatives of the species that we have already domesticated. Natural ecosystems are also a source of raw materials. Carefully managed semi-natural forests can provide timber indefinitely, while also conserving much of their original biodiversity; mangroves and coral reefs act as the nursery grounds for a significant proportion of the fish that the world consumes.

### MEDICINE

Tropical rainforests are prolific sources of plants with medicinal values. Although there is a vast potential for improving human health using the services provided by

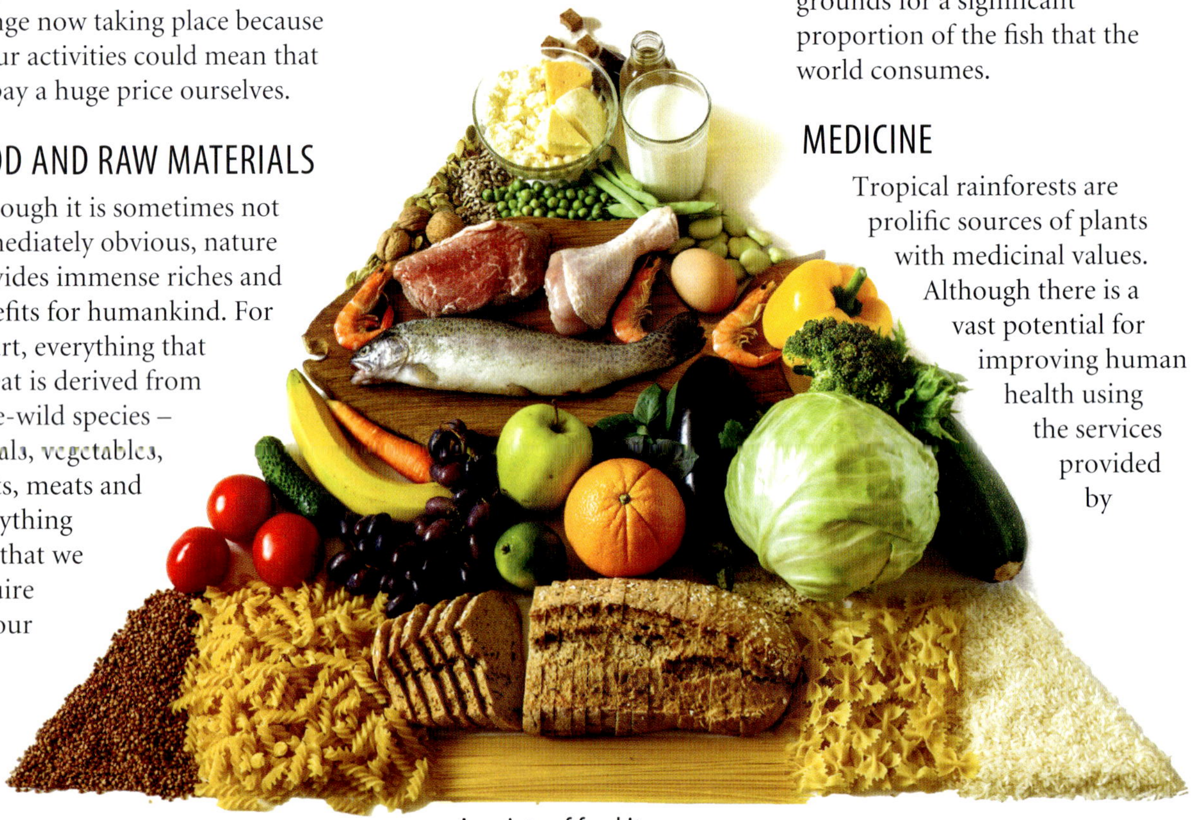

A variety of food items

*The smoke, poisonous emissions, industrial waste, and contamination of marine biomes are harming all the living beings and also strangling Mother Earth*

Amazonian frog

biodiversity, we are wrecking that potential before we have barely begun to understand it. For example, more than 2,000 tropical forest plants have been identified as having some form of anti-cancer properties. But, of these, we have only tested one in ten for their anti-cancer properties and only intensively screened about one per cent of them. Many are going extinct before we have even assigned them names, forget about assessing their potential to help us. Tropical rainforest animals have also made contributions to medicine. For example, a species of Amazonian frog has contributed a compound that helps in the treatment of strokes, seizures, depression, and Alzheimer's disease.

The tropical rainforests act as a global air conditioner, absorbing and storing $CO_2$ from the air and releasing oxygen. They also play a major part in regional water cycles, generating rain and sustaining the flow of rivers.

## Did You Know

Malaria, has for decades been treated with quinine, first found in the bark of the cinchona tree. The muscle relaxant curare, which is used during surgery, was first extracted from a species of a vine.

Cinchona tree

*Activities related to ecotourism such as backpacking and hiking, control the negative impacts on the natural environment*

## TOURISM AND THE FIGHT AGAINST POVERTY

International tourism is a big grosser for foreign currency and an important factor in the balance of payments of many countries. A large chunk of this economic activity is driven by people wishing to spend time with

*Also called the happy tree or cancer tree, Camptotheca's extracts were used to develop the chemotherapeutic drug Topotecan*

*Encouraging the younger generation to plant seedlings and care for a tree helps them learn the importance of natural world*

nature: bird-watching, trekking and reef diving, among other activities. The destruction of these assets is, therefore, very likely to lead to reduced future revenues in a vital sector.

Poorer societies depend directly on the natural resources for their well-being: on productive soils, reliable water supplies, prolific fisheries, wild meat and timber from natural forests, among other things. As 'development' pressures destroy or degrade these services and assets through unsustainable patterns of land use and resource management, the prospects for human development decline rather than improve.

## OUR MORAL RESPONSIBILITY

We also have a moral responsibility to conserve the Earth's natural diversity. The Earth's great natural riches have emerged following three and a half billion years of evolutionary change. Does any single species have the right to destroy much, or even most, of that unique planetary heritage? Our present actions seem all the more indefensible given that what we are now doing is with our full knowledge.

## HOPE FOR THE FUTURE

Fortunately, all is not yet lost. It is possible for us to take steps now that will enable us to pass on intact much of the natural world and the irreplaceable services it provides for us. What is needed is collaboration with local communities to achieve large-scale habitat conservation, putting pollution control measures in place, proper enforcement of species protection, augmented, where appropriate, with captive breeding programmes and the control of introduced animals and plants.

## BACK FROM THE GRAVE

Some creatures that were thought to have been extinct – at times for millions of years – have sometimes miraculously been rediscovered again in modern times, proving that when given a chance, life finds a way to survive.

## CONSERVATION ACTIONS UNDERWAY

The government is taking a number of steps to conserve the species. The help of the local Yanaadi community, members of which played a big role in its rediscovery, has been sought by the State Forest Department for locating it in other habitats. The Lankamalai and the Veliconda hill range have been declared as wildlife sanctuaries. Also, the Telagu-Ganga canal, which would have passed through one of these protected areas, has been realigned to prevent fragmentation of the bird's habitat.

### Did You Know

The Coelacanth are an ancient order of fish believed to have gone extinct with the dinosaurs 65 million years ago. That was until 1938, when one was discovered off the east coast of South Africa near the mouth of the Chalumna River. There are only two known species of coelacanths: one that lives near the Comoros Islands off the east coast of Africa, and one found in the waters off Sulawesi, Indonesia.

West Indian Ocean coelacanth

## Did You Know

The Jerdon's courser was first sighted by the surgeon-naturalist Thomas C Jerdon in the Pennar and Godavari river valleys in 1848. It later went missing for almost a century until its rediscovery in 1986 in the Lankamalai area of Andhra Pradesh.

Jerdon's courser

The La Palma giant lizard was believed to have been extinct for around 500 years till its rediscovery in 2007. Though only one individual was sighted in the La Palma region of the Canary Islands, its presence suggests that a few of them might still exist.

While we are encouraged by these chance rediscoveries, we still need to take steps to preserve, protect, nurture and sustain the biodiversity that makes our world a unique place. To do so, we must be clear about priorities and the measures needed for conservation, locally, nationally and internationally.

## PROTECTED AREAS

The first step to prevent loss of biodiversity is to secure the remaining natural habitats. For example, Madagascar is home to 10,000 species of plants, 80 per cent of which are found nowhere else in the world. Incredibly, in 2006 alone, some nineteen new species of primates were described for only the first time. But deforestation on the island is placing animals, such as the lemurs, at risk. Of course, efforts are on to encourage some system for protection of habitat, such as setting up a network of national parks in most countries, and there are more than 100,000 protected areas worldwide, covering almost 19 million square kilometres. However, unlike the Great Barrier Reef Marine Park in Australia,

*Every year, 1.6 million tourists visit the Great Barrier Reef Marine Park for diving and boating excursions. The park is actually a conservation effort, saving the Great Barrier Reef – one of the seven wonders of the underwater and natural worlds*

*Once thought to be extinct, the La Palma giant lizard occurs only in the Canary Islands*

and the Amazonia forest reserve in Colombia, most of the Wildlife Reserves are fragmented patches. Additionally, less than 10 per cent of the world's major lakes are part of the protection plan. In the marine environment, the situation is far worse, as less than 0.5% area is protected.

Another worrying thought is that many threatened species do not occur within any protected area, leaving hundreds of endangered species vulnerable. It is vital that the protected areas are managed in the correct way. Habitat diversity can be increased

through various kinds of human use. For example, in some nature reserves, access must be tightly controlled and activities such as logging and hunting forbidden. Other habitats, which are fragmented, may actually require human activities such as grazing or water level management to create conditions that would only occur naturally in larger patches of natural habitat. Thus, to retain biodiversity, it is necessary to not only protect the remaining habitats but also to recreate, restore and expand those that are lost, fragmented or degraded.

## RESTORING THE EARTH

Programmes are in progress across the globe to restore and recreate natural habitats. In fact, these efforts are already paying dividends. Conservation groups, official agencies and volunteers are restoring habitats from the degraded fenlands of eastern England to the Salt River near Phoenix, Arizona, from the heavily degraded and polluted Chesapeake Bay on the east coast of the US to the fragmented Atlantic rainforest of eastern Brazil.

*The Chesapeake Bay – a complex ecosystem, including important habitats and food webs – is the largest estuary in the United States*

## PEOPLE AS PARTNERS

Maintaining biodiversity of protected areas and habitat restoration can work well only when local people work as partners in the conservation programmes. Human beings have often been seen as intruders who damage biodiversity and its conservation programmes. This view is quite wrong. Many habitats that are today regarded as 'natural' have, for long periods, been shaped by the activities of humans. For example, the native people living in Brazilian Amazonia were not just passive inhabitants of this complex ecosystem. They selectively burned small patches of land to grow crops such as manioc and sweet potatoes. When these primary crops were harvested, fruit trees were planted, not only providing a further crop of food but also attracting fruit-eating animals. The pattern was repeated in other areas, providing secure food supplies, while at the same time sustaining the forest. Over long periods, this shifting cultivation subtly changed the ecosystem in ways that were often beneficial to wildlife.

*Home to the most diverse set of species on Earth, the Yasuní Biosphere Reserve is part of the Amazon Basin*

Like the wildlife, native people and other tribal communities, who can be powerful allies in sustaining the Earth's natural riches, are being wiped out. By the year 2000, there were less than 250,000 native people still living in the Amazon Basin as they had been killed for their land, or pressed into slavery, or again had died from introduced diseases to which they had no resistance.

These local people can be a very powerful and effective conservation strategy. In fact, the present-day map of rainforests

*The Amazon rainforest in Brazil is not only home to rare species of animals and birds, but indigenous people and tribes as well, who remain isolated even from other indigenous people*

in the Amazon basin shows that the areas of intact forest are those where the lands are still controlled by the native forest inhabitants. Their culture and lifestyles are far more favourable to sustaining the forests than that of the newcomers who ravage the natural resources of the forests.

## REDUCING POLLUTION

While controlling excessive fishing in coral reefs is essential for conserving marine biodiversity, this is insufficient on its own. To maintain their diversity, coral reefs need to be protected from different kinds of pollution. Development programmes taking place at a distance from these reefs also need to ensure reefs are not harmed by making new tourist hotels install modern sewage treatment facilities. Use of chemical pesticides and fertilizers in agriculture also needs to be limited to check damage to these reefs.

*Classified as Endangered by the IUCN, many blue whales get injured or die each year from impacts with large ships*

### Did You Know

The International Convention for the Regulation of Whaling was signed in 1946 to effectively manage the hunting of whales. The agreement led to the establishment of the International Whaling Commission, to "provide for the proper conservation of whale stocks and thus make possible the orderly development of the whaling industry". The Convention includes a legally binding Schedule which, amongst other things, sets out catch limits for commercial and aboriginal subsistence whaling.

## PROTECTING THE RAREST

Martha, the passenger pigeon, did not survive long enough to see the passing of new laws protecting rare and declining species animals and plants from extinction.

There are also a number of international rules and laws like the Convention on the International Trade in Endangered Species (CITES) that have been signed and ratified by nearly all countries, wherein certain listed animals and plants are banned from international trade. There are problems with this agreement – for example, lack of available resources in some countries for them to enforce it effectively – but it has undoubtedly saved many species from being driven to extinction.

*Dugong lying on seabed*

Regulations and policies requiring industry to reduce or end the use of certain harmful substances have decreased pollution levels, reducing pressure on wildlife populations.

DDT has already been banned from farm use in many countries following the disappearance of predatory birds, leading to the recovery of severely depleted bird populations.

*The remaining populations of the Cambodian crocodile in the wild only appear in the Mekong River basin and wetlands in Cambodia*

*In 2005, due to the conservation efforts of the Nature Conservation Centre, Al Ain, there has been a considerable increase in the number of Arabian tahrs*

But laws and treaties are only as good as their enforcement. The financial rewards from illegal hunting tempt many to break the law. Another protected yet critically endangered species that is threatened with extinction through illegal hunting is  the Cambodian crocodile. These creatures are highly sought after by crocodile poachers throughout Asia: an adult crocodile fetches up to $2,000 in the black market. As a result, only around 250 adult Siamese crocodiles remain in the wild, chiefly in the remotest highlands of Cambodia, .

What is needed is not just strong enforcement backed by tough penalties but also measures to educate buyers and to discourage the creation of markets in rare species in the first place.

## BREEDING AND RELEASE

When a species declines to the point at which it is nearly lost, or indeed has already become extinct in the wild, then captive breeding can be a useful means to aid its recovery. Nearly a fifth of all mammals and one tenth of bird species have been successfully bred in captivity. The population of Arabian tahrs in UAE and red wolves in the USA has increased due to captive breeding. Conservation through captive breeding reduces genetic diversity. As a result, zoo-bred animals lack immunity to disease and disease outbreaks can have disastrous consequences. Being kept in captivity weakens the animals' ability to survive in the wild, and the animals are also often unable to re-adapt to their new environments. Moreover, studies have shown that once released, their reproduction rates plummet. Captive breeding also creates a

Red wolf

**Did You Know**

The lion-tailed macaque, indigenous to the tropical rainforests of the Western Ghats in Southwest India, is one of the most critically endangered species with less than 2,500 individuals left in the wild. In order to safeguard the species, they are now being bred in captivity.

A lion-tailed macaque

Cane toad

false sense that the battle to save endangered species and habitats is being won.

Therefore, while the importance of captive breeding as a recovery strategy cannot be discounted, it is essential that natural habitats be preserved and recreated at the same time as animals are reared, so that they can be subsequently released in nature.

## CONTROLLING AGGRESSIVE INVADERS

Alien species have caused widespread damage wherever they have been introduced. It is difficult to totally eradicate established invaders as their elimination requires long-term commitment. Failure to be effective in controlling introduced species could lead to serious global consequences as humans, agricultural output, forestry

*Red swamp crayfish and cane toads are invasive species that can grow fast by predating, thereby competing for food resources with native populations and affecting their growth*

and fisheries are affected by these invaders. While centuries ago people introduced exotic creatures in new areas in their ignorance, today we know a great deal about their potentially devastating impact and should make every effort we can to remove the ones that have already become established to the detriment of native animals and plants.

## GREENHOUSE EFFECT

There is, however, a still bigger threat. The biosphere does not only interact with itself. It fundamentally also relies on the functioning of the atmosphere. And the atmosphere is changing, and changing fast. When the Earth was formed about 4.5 BYA, its early atmosphere was probably mostly $CO_2$, with little or no oxygen. There were smaller proportions of water vapour, ammonia and methane. As the Earth cooled down, most of the water vapour condensed and formed the oceans. The proportion of $CO_2$ in the atmosphere went down, and the proportion of oxygen went up because of photosynthesis by plants.

Today, the burning of fossil fuels is adding $CO_2$ to the atmosphere faster than it can be removed. This means that the level of $CO_2$ in the atmosphere is increasing.

The rising human population is adding to atmospheric $CO_2$ in other ways too. Deforestation leaves fewer trees to remove $CO_2$ from the atmosphere by

### Did You Know

In temperate regions, the winters are getting warmer, spring is arriving sooner and summers last longer. This will increase the reproduction rate of insects, which will in turn prove harmful for the crops.

Colorado potato beetle larvae eating the foliage of potato

*A team from the Norwegian Polar Institute documented polar bears eating dolphins for the first time and subsequently freezing the leftovers under ice for future needs*

*Considering the present environmental circumstances, only smart, green cities seem to be holding the key to future power needs and a greener Earth*

photosynthesis. Moreover, if the fallen trees are burned or left to rot, additional $CO_2$ is released into the atmosphere, further adding to the problem. As $CO_2$ levels rise, the heat of the Sun is trapped by the atmosphere, thereby increasing the temperatures. If the Earth's temperature gets hotter by just a few degrees, it could change the weather all over the planet in big ways. As places that grow most of our food get too hot to grow crops anymore and ecosystems start changing, some species will move farther north or become more successful; others won't be able to move and could become extinct. Wildlife research scientist Martyn Obbard has found that since the mid-1980s, with less ice on which to live and fish for food, polar bears have gotten considerably skinnier. He fears that if sea ice disappears, the polar bears will disappear too.

The tropical insects would start dying out as they are unable to adapt to a warmer climate. It would harm the plants they pollinate and the animals the plants, and the insects feed on.

## TIME TO SAVE THE EARTH

The heartening thing is that there is a growing awareness worldwide that the Earth is a living organism, of which we are parts. It needs to be respected and preserved. The World Commission on Environment and Development put forward the concept of sustainable development in 1987. In its report, it defined the idea as "Development that meets the needs of the present, without compromising the ability of future generations to meet their needs." This emerging new world vision has ushered in the Era of Responsibility. We must remember the statement made by Margaret Thatcher, former Prime Minister of the United Kingdom, "No generation has a freehold on this Earth. All we have is a life tenancy – with a full repairing lease". In the words of Mr Lester Brown, "We have not inherited this Earth from our forefathers; we have borrowed it from our children."

# GLOSSARY

- **Ammonoids** – an extinct group of marine mollusc animals in the subclass Ammonoidea of the class Cephalopoda
- **Amphibians** – group of vertebrate animals characterized by their ability to live both in water and on land
- **Arthropods** – (e.g. crabs, insects, spiders) are unique among invertebrates in that they have clearly separate senses of taste and olfaction that are comparable to those of vertebrates
- **Asteroids** – a large number of small bodies, about 1,000 km (600 miles) or less in diameter, that orbit the Sun
- **BCE** – Before the common era
- **Cnidarian** – invertebrate also called coelenterate any member of the phylum Cnidaria (Coelenterata), a group made up of more than 9,000 living species
- **Comet** – a small body orbiting the Sun made up of volatile ices. When a comet comes close to the Sun, the ices change to gas and along with entrained dust particles form a bright outflowing atmosphere around the comet nucleus known as a coma
- **Cycads** – a diverse collection of mostly extinct primitive gymnospermous plants
- **Fossil** – paleontology remnant, impression, or trace of an animal or plant of a past geologic age that has been preserved in Earth's crust
- **Gymnosperms** – represents four surviving divisions of vascular plants whose ovules (seeds) are exposed on the surface of cone scales
- **Hominid** – a primate of a family (Hominidae) which includes humans and their fossil ancestors and also (in recent schemes) at least some of the great apes
- **Hominin** – a primate of a taxonomic tribe (Hominini), which comprises those species regarded as human, directly ancestral to humans, or very closely related to humans
- **Isthmus** – a narrow strip of land connecting two large land areas otherwise separated by bodies of water.
- **Nautiloids** – a large and diverse group of marine cephalopods belonging to the subclass Nautiloidea that began in the Late Cambrian and are represented today by the living Nautilus and Allonautilus
- **Nebula** – "mist" or "cloud" of gas and dust that occur in interstellar space
- **Photosynthesis** – the process by which higher plants manufacture dry matter through the aid of chlorophyll pigment, which uses solar energy to produce carbohydrates out of water and carbon dioxide.
- **Phytoplankton** – a plant species of freely floating, often minute organisms that drift with water currents
- **Radiocarbon dating** – method of age determination that depends upon the decay of radiocarbon (carbon-14) to nitrogen
- **Sassafras** – a tree of the laurel family
- **Sedimentary rock** – rocks composed of particles, or grains, of weathered rock or mineral debris
- **Solstice** – either of the two times in the year, the summer solstice and the winter solstice, when the sun reaches its highest or lowest point in the sky at noon, marked by the longest and shortest days
- **Stadia** an ancient Roman or Greek unit of length, about 185 m (originally the length of a stadium)
- **Strait** – a naturally formed, narrow, typically navigable waterway that connects two larger bodies of water
- **Supernova** – a class of violently exploding stars, whose luminosity suddenly increases many millions of times its normal level after eruption
- **Tetrapods** – a superclass of animals that includes all limbed vertebrates (backboned animals) constituting the classes Amphibia (amphibians), Reptilia (reptiles), Aves (birds), Mammalia (mammals), and their direct ancestors that emerged roughly 397 MYA during the Devonian Period
- **Waning** – phase of the moon when it appears to become thinner or less full
- **Waxing** – phase of the moon when it appears to become larger or more full